Once an Engineer

SUNY series in Italian/American Culture
Fred L. Gardaphe, editor

Once an Engineer
A Song of the Salt City

Joe Amato

excelsior editions

Cover photo credit: 501 *Raphael Ave* by Michael Amato.

Published by State University of New York Press, Albany

© 2009 Joe Amato

All rights reserved

Printed in the United States of America

No part of this book may be used or reproduced in any manner whatsoever without written permission. No part of this book may be stored in a retrieval system or transmitted in any form or by any means including electronic, electrostatic, magnetic tape, mechanical, photocopying, recording, or otherwise without the prior permission in writing of the publisher.

Excelsior Editions is an imprint of State University of New York Press

For information, contact State University of New York Press, Albany, NY
www.sunypress.edu

Production by Ryan Morris
Marketing by Fran Keneston

Library of Congress Cataloging-in-Publication Data

Amato, Joe, 1955—
　Once an engineer : a song of the Salt City / Joe Amato.
　　　p. cm. – (SUNY series in Italian/American culture)
　Includes bibliographical references.
　ISBN 978-1-4384-2843-7 (hardcover : alk. paper)
　1. Amato, Joe, 1955—Childhood and youth—Anecdotes. 2. Syracuse (N.Y.)—Social life and customs—20th centiury—Anecdotes. 3. Amato family–Anecdotes. I. Title.

F129.S8A437 2009
974.7'66043092–dc22
[B]

2009003932

10 9 8 7 6 5 4 3 2 1

for memory, and against

And herein lies the tragedy of the age: not that men are poor,—
all men know something of poverty; not that men are wicked,—
who is good? Not that men are ignorant,—what is Truth? Nay,
but that men know so little of men.
— W. E. B. Du Bois, *The Souls of Black Folk*

Kee vah sah-noh, vah pee-ah-noh
Ay kee vah pee-ah-noh
Vah lohn-tah-noh.
— Rosario Amato, transcribed ca. 1974

I would have the engineers join in the drinking and the dancing.
— Samuel C. Florman,
The Existential Pleasures of Engineering

Contents

Part I — Bildung

1	Winter Rat	3
2	Landscaping	13
3	Wicked Piss	31
4	Games People Play	47
5	The Flying Pork Chops and Other Adventures in Craft and Cuisine	67
6	Linkage	87

Part II — Rebuilding

7	Salt City	105
8	Just Produce: A Meditation on Time & Materials, Past & Present	127
9	Primitive Roots	147
10	Say-Cursed Susan B. Anthonies	159
11	Notes toward a Supreme Fiction	185
	Epilogue: Variable Cloudiness, Chance of Precipitation 50%	251
	Acknowledgments	259

Part I
Bildung

I.

Winter Rat

> You can observe a lot by watching.
> —Lawrence Peter Berra

IN CENTRAL NEW YORK, there are two kinds of winter rats.

First kind: the car you drive winters only. If you can manage it. You own two vehicles, and take your so-called first vehicle off the road. So-called because snow can generally accumulate anytime between 1 November and 1 May. Which means that for six months of the year, you're driving a shitbox. And this is six months of rough winter. Few populated areas in the U.S. get the sort of snow you see around Syracuse.

Lake-effect snow: in the Tug Hill plateau, a narrow stretch of land an hour's drive to the north, as much as three hundred fifty inches a season. In Syracuse itself, as much as two hundred. In the zone between, it varies. A slight change in wind pattern can be the difference between six inches and three feet. (The record for the region, set during the 1976–77 season, is just under thirty-nine *feet*.) As kids on snowy winter mornings, we get up early and turn on the radio first thing, listen intently to hear if school will be closed due to the weather. The first closing is always Altmar-Parish-Williamstown Central, up near the Tug Hill region. But the Liverpool School District has a good bus system, rarely shuts down.

Living where we do, on the edge of the Salt City, you can expect ten feet of snow anyway each winter, what the local news networks refer to as "hard-packed snow and ice." And slush and hail and sleet and rain. And no shortage of salt and sand to help the driving conditions. So driving a shitbox, a crate, a rat for these six months represents a real aversion to rust. A real commitment to car culture.

It's a custom the three of us can't afford. We drive my high school graduation present—a three-year-old 1969 Camaro Rally Sport with hideaway headlights, a gift from my mother—year-round, until the accident. Then it's an endless series of shitboxes.

Like I say, winters are rough, unending, pretty only when the snow is new. And as they say, misery loves company. A kick in the ass. So an unpredictable climate is accompanied by a predictable affectation you find only in Central New Yorkers.

Weather SUCKS—worst on the planet. They have less snow in Anchorage. But it's a great place to live, FUCK YOU.

Central New Yorkers agree—the purpose of winter is to make you work to be happy. We're proud to be so happy in our misery. And to add to our miserable happiness, if not our pride: assholes who don't know how to drive in the snow.

Early morning, still dark. Barely conscious, I can hear somebody gunning an engine, then letting up, then gunning it. It's Freddy, he's stuck. Again.

Asshole.

Joey c'mon get up. Let's go.

I sit up, reach over and pull on my pants, standing. Mike is still sleeping. I stumble out into the kitchen. Pull on boots, no socks. Throw a coat on over my white tee. My father, Parliament glowing in one hand, hands me a pair of gloves. We walk downstairs, I brace myself as we walk out into the cold. The wind is blowing snow up into my glasses. My father holds his cigarette cupped in his hand. Asshole's car sits halfway into the road, halfway out of the driveway, one rear wheel spinning helplessly.

Geez. Hi Joe. Thanks.

My father puts his cigarette to his lips. We get behind and push Asshole. It takes about twenty seconds to push him through the pile of snow that the plow has left at our dead end, square in front of our so-called driveway. Sometimes the guy driving the plow understands and turns his blade, shoving the snow up and past. Sometimes he doesn't. If he doesn't, Asshole gets stuck, because Asshole babies his shitbox through the snowbank. And when Asshole gets stuck, we're stuck behind Asshole.

My father coughs a couple of times, catches his breath. Then puffs on his cigarette. Freddy drives off, sliding this way and that, the way now clear for us. I turn and stumble back upstairs, catch a little shut-eye before I have to get up.

Winters are just *made* for assholes.

Second kind of winter rat: a rat in winter.

Joe, wake up. I heard something.

Go to sleep.

Winter Rat

We're sleeping in the same bedroom, on single beds a few feet apart. At the foot of my brother's bed sits a baby blue, beat-up dresser, with white trim. I turn over, close my eyes.

Joe—wake up, I heard something.
Go to sleep.
But I heard something.
Where?
In the closet.
You're full of shit.
You're full of shit—I heard something.
By now we're getting loud. Mike gets up, turns on the light.
Look!
What?
I saw it!

I sit up. I hear the click of a lighter. My father enters, wearing boxer shorts and socks, a Parliament in his left hand. No shirt. I'm sitting up in bed now. All three of us are in our underwear.

What's going on boys?
Mike says he saw something.
A rat!
A rat?
Dad—I saw it—a rat—it ran in the closet!
Shit. OK—get out of the way.

My father pushes my brother's bed back a bit, out of the way. Mike sits on the bed, I stand—we're both peering over his shoulder. Kneeling and reaching in, he begins, slowly, to remove each item from the closet. Three baseball bats, my metal tackle box, Mike's magic stand, my Jon Gnagy art kit. A small cloud of dust follows each item my father touches. He's using one hand, then the other, alternately holding and puffing on his cigarette. The underside of the attic steps comprises the closet's ceiling, which tapers to eight inches high at its far end.

His hands shake a bit, but not because he's nervous. He carefully inspects each box, each item. Nobody says a word, just an occasional huh, breathing.

As he nears the back of the closet, all that remains are three rolled-up posters. These I've had since childhood. I bought them mail order through an ad in the back of one of my *Famous Monsters of Filmland* mags. One poster is a full-size, garishly hued rendering of Dracula, another a similar rendering of Frankenstein's monster, the third a large black & white close-up of Lon Chaney Jr.'s head as the Wolf Man.

My father picks up the first poster, tilting it sideways. Nothing. He hands it to me, I unroll it a bit. Dracula.

He picks up the second poster, tilting it. Nothing. Hands it to me again. Frankenstein.

Is Mike dreaming?

My father pauses, puts his cigarette to his lips, and puffs, letting the smoke out with the cigarette still in his mouth. Then he leans in slowly, balancing his body on his free hand, reaching with the other to pick up the final poster. His left hand closes gently around the curve of the rolled cardboard. Gingerly, warily, he pulls the poster out of the closet, tilting it slightly toward him.

Out leaps an eight-inch-long RAT.

All three of us gasp. My father drops the poster, falling backward and smashing up against the bed, his cigarette dropping out of his mouth. Mike does a backward roll off the bed onto the floor, landing on his feet, I leap up on the dresser. The rat tears ass across the bedroom, slipping across the floor out into the kitchen. And under the stove.

Jesus-Christ-All-Mighty!

I grab one of the baseball bats my father has removed from the closet. Mike and I lace up our sneaks. My father puts on his leather shoes and a short-sleeved button shirt. We're all three of us still in our underwear.

We close the door leading to the bedrooms. We put up a barricade of boxes between the kitchen and the living room.

Mike arms himself with this two-foot-long, spring-ended shoehorn—a souvenir from the Adirondacks my father has never used, but keeps hanging on a nail in his bedroom.

Give that to me!

My father grabs the shoehorn from Mike.

I'll fix his ass.

My father turns on the stove and oven, and we wait. The oven pops slightly as it expands from the heat.

Suddenly we hear the rat moving.

Getting hot under there.

It darts out, behind the refrigerator. A moment later, it runs to the corner of the kitchen, near the entrance, behind my father's finishing box, a box full of aniline stains, lacquer sticks, putty knives, irons, sanding blocks. Sixty pounds' worth. On top of which is a stack of newspapers, grocery bags, and more tools.

We figure we've got the rat cornered now. My father tells us to stand back. I'm sitting on the counter, Mike is standing near the stove. Again, my father picks up, slowly, each item on top of the finishing box. Nothing. The final item is the box itself. My father pauses for a moment, and with some effort grabs a hold of it and swings it out of the way.

The rat is there all right—and cornered. It gets up on its hind legs and lets out a screech, leaping a full foot-and-a-half into the air at my father's face. My father jumps back, falling out of the way, and the rat tries to scamper past him. My father turns and, jerking the shoehorn up like a flyswatter, lets it down with a WHACK, catching the rat mid-back. It collapses, twitching, and my father jerks and lets the shoehorn down again. Fucking WHACK. Red spills out of the rat's mouth all over the kitchen floor.

A moment later, my father is pushing the rat onto some newspaper with the shoehorn. He takes it outside. Then he cleans up the floor with a paper towel and Windex.

South Dolores Terrace is a white working-class suburb, which is to say a lower-middle-class, fifties-style suburb, on the poor side of *Leave It to Beaver*. The three-bedroom ranch houses cost a little over fifteen grand in 1956. Only a couple of the fathers are salary workers, hold college degrees, thanks to the GI Bill. And only a couple of the mothers have any college at all.

My father has lived through the Depression, in an immigrant family with four sons. He's picked up his high school equivalency after the war, with my mother's help. He's strictly hourly, union, a union steward for a spell.

But furniture finishing has never itself been unionized. So when General Electric leaves him—or he leaves General Electric—he's left without a trade union. And ends up working for entrepreneurs. You can think of them as the bourgeoisie, if you like. Me, being the son of a French woman, I take the former term literally, and think of them as undertakers.

Boys, I've never been this poor in my life.

Then, as now, many of the poor own TV sets. They're even easier to come by than during the fifties and early sixties. My father was employed at General Electric for nearly twenty years as a touch-up man. He worked on an assembly line dogleg, fixing up nicked and damaged wood TV cabinets. Back when TV cabinets were furniture.

On 112 South Dolores Terrace, we've always got a couple of older tube units around. My father's friend, Johnny Palamino, runs a TV repair shop on the north side, and throws cabinet work my father's way. After the divorce, at 501 Raphael Ave. now and thanks to Mr. Palamino, my father always manages to find a decent used set, hooking up a good pair of rabbit ears for a so-so picture. If we're lucky, we get a color set. It usually lasts five or six months. Then we get a new used one.

During those hours my father, brother, and I spend together in front of the TV, my father chain smokes, or nearly so. Anywhere from two to five packs a day, depending on how nervous he is. Which depends on the situation that day. Ditto for alcohol consumption. The more suspenseful the film, the more smoke, the more drink.

And at night, my father falls asleep in front of the box, volume turned down low. Sometimes, in the early morning hours, long after the channels have signed-off for the night, I get out of bed to turn off the test pattern.

Joe—whada?—leave it on!

But Dad—

DO AS I SAY, and make it snappy!

A lasting image, years prior to my reading of Orwell: my mother, light-tanned complexion and auburn hair, relaxing on the front concrete steps one sunny afternoon at 112 South Dolores Terrace, admiring her rosebushes. This is back when we still have a septic tank, when our driveway still supplies pebbles for my rock collection. Fossil days.

Suddenly my mother lets out a scream.

My father runs out of the garage to see a large rat running, in broad daylight, across the front yard. Instinctively he reaches for a garden rake—one with short, pointed steel tongs. He chases the rat across the street, rake held high over his shoulder like a madman, spiking the rat to our neighbor's lawn.

Turns out the creature had burrowed under one of the bushes along the front of our house. On Dolores Terrace, no rats permitted. The exception proves the rule.

But that rat at 501 Raphael Ave.—which locals pronounce *Ray-feel*—is really just the beginning. Or beginning's end.

For Mike, the attic becomes an arena in which to test his survival skills. As a kid, he's a tree-climber, model rocket builder, GI Joe collector, BB gun owner. Years later, it's a .357 mag, and he's an expert rock climber, hikes the Grant Teton summit without any technical gear. But at 112 South Dolores Terrace, he straps on the camouflage mock-parachute he's picked up for two bucks downtown, at the Army-Navy store on Clinton Street, and jumps off the roof of our house, hoping the chute will break his fall to earth. The hard way.

Thirty years later, he'll buy and sell government surplus—online—for a small fortune.

Crow, squirrel, raccoon haven, dark steps ascending. Moving into 501 Raphael Ave., we abandon to the attic the remains of Dolores Terrace. Sleeping, we hear noises in the walls. And in the kitchen, as luck would have it, large ants attack, depending on the season—but no roaches.

Joists left open in the attic. After the rat, we hunt around. There, my seven-hundred comic books, yellowing. Here, a suitcase full of letters, photos, drawings, clothing, cloth—from Europe, stuff my mother must have overlooked. There, a box of toys, most in pieces. We've ripped the heads off of our monster models, save them in a cardboard box as mementos—of what I don't know.

One evening, Mike and I hear a commotion. We climb up the steps together, slowly. We push open the attic door and flip it back on its hinges. A heavy wooden item. Flip it up hard or it'll flip back and drop on your head. As we peer out into the darkness, waist high into the attic, a large shadow comes swooping directly at our heads, squawking. We duck and drop down the steps, the attic door smashing over the opening.

What the hell was that?

We're trembling now—this is fun. We creep up the steps again, flip up the door. Make fast for the light bulb—pull the string, and we can see a giant crow making directly for us. We duck, it swoops over. We run back downstairs. The door smashes down again.

My father goes up, manages to open a window.

Mike has taken to buying mousetraps. Then small animal traps.

Now some of these creatures are crafty. The cheese is nibbled off the trap come morning. But Mike is craftier still.

He gets a bucket, fills it with water six inches shy of the top. Then he takes a flat stick, and leans it over the edge of the bucket, forming a sort of ramp. Next, he creates a path of cheese morsels, leading up the ramp. Finally, he dangles a cheese morsel from a string attached to the attic ceiling, dead center over the bucket and perhaps six inches from the end of the ramp.

Next morning, a dead mouse is floating in the bucket. Its small furry body tits up and twisted, a tiny bit of shit in the water.

On a Saturday morning when my father has gone on a house call, Mike and I awaken to a faint but rhythmic high-pitched PINGPING coming from the kitchen. We both get out of bed, sneaking over to the doorway that separates the bedrooms from the kitchen. PINGPING. The noise is coming from inside the top portion of the stove, under the burners. From the doorway, we can both twist our heads around the corner to peer in behind the stove. PINGPING.

Mike gives the stove a push, and a mouse darts out of it and down the gas piping, through the small chewed opening just above where the piping pokes through the drywall.

We'll get you, you sonofabitch.

The next morning we hear the PINGPING again. We sneak over to the doorway. PINGPING. I hand Mike a firecracker, he strikes a match and lights the fuse, at the last possible instant tossing it inside the stove.

BA-BAM!

The mouse, shell-shocked, nevertheless makes it back down the pipe and through the opening. When my father gets home, we tell him about it. He thinks it over for a moment.

You know, you boys should be careful blowing off firecrackers around gas piping.

That day my father decides it's best to seal up the opening, leave the other side to the rodents.

Up in the attic again, Mike sets a large mousetrap. Large enough to break your hand. One night, sleeping, we're awakened by a loud SNAP coming from above the ceiling, followed by a number of dull thuds.

The next morning, we survey the scene: there's the trap, sprung, and a few feet away is a large, black, dead squirrel, with an injury to the head. Its mouth crooked open, its teeth large and pointed. My father shakes his head.

Meeng-kya! Good work, Mike.

My father walks downstairs, comes back up a moment later with a pair of gas pliers. He grabs the squirrel with the pliers—gently, or so it seems to me—and carries it over to the half-open attic window, tossing it out with little ado. It lands on the concrete walk to what used to be the front entrance.

Months later, looking down from the attic window, one can see the skeleton of the squirrel. Years later, its faint outline still graces the concrete. Whenever we look down at it, we chuckle.

Even I get into the act a bit.

A short while after we rearrange the flat so that Mike and I can have our own bedrooms, and for several weeks, I'm noticing small bumps on my chest. They itch, and they last for a few days. And new bumps seem to appear between evening and morning.

One night, late, I'm awakened by the sensation that something is biting me. I jump up out of bed, turn on the bedroom light, and get a flashlight. Mike wakes up in the adjoining room, laughing.

What are you doin'?

Bites.

He laughs harder.

I duck under my bed, flashlight in hand, waving the beam across the floor and walls. Dust everywhere. In the corner, I notice the beam glisten on what looks to be a cobweb. But pointing the beam directly at the web, I can just make out a small spider, crouched in the corner. I reach under the bed and swat at it, crushing the spider with the flashlight.

Now my mother always says that killing a spider is bad luck. But my first comic book as a small child, the prehistoric caveman-warrior Kona, is all about a giant spider that's been trapping T. rexes, brontosaurs, and other dinosaurs in its huge web. Ever since then, except perhaps for daddy longlegs and a few of the tinier varieties, I don't like spiders.

I notice another small web, a few feet further along the wall. Again I spot the spider, and kill it with the flashlight. A few feet more, another spider—and another dead spider. And another. And another.

That night I locate and kill sixteen spiders in all—some small, some not so small. The bumps vanish in a week.

Turning into the neighborhood late one August night, driving the Camaro, Mike and I catch a glimpse of a shadow hobbling along the front of the landlord's plumbing and heating supply company. I stop the car, back up, and pull up to the building, shining my headlights at it.

It's a large RAT, perhaps ten inches in length, perhaps eight inches in girth.

A winter rat may double as a summer rat, you see—there are rats for all seasons. You have to learn to live with such things.

We laugh it off. Or yell at each other.

We dream. Or turn on the TV. We tune in the radio, or play a record.

Whatever's available, cheap. A good horror film, say, to remind us of what we could be, animal man and woman alike. Are.

My father's father, Rosario—he learned how to laugh early on, I suspect. I imagine him, a new immigrant, muttering under his breath all sorts of funny things in Italian.

Maybe, not so funny.

But immigrant or no, he was a man, finally, in a man's world—a world where men enjoy the greater liberty of their lesser natures.

That saying of my grampa's, the one with which I begin this book—proverbially, *Chi va piano va sano e va lontano*. But it's not quite "slow and steady wins the race," as some would have it. As my grampa had it, it was more about going far—*lasting*.

And when we've finished, we brawny mortals are finished from first to last.

Meanwhile men and women will do what they have to to get by. Sometimes it's not a pretty sight. It takes gall, for one, which might or might not translate into galleys, gallows humor, Gallicisms. That's my job, *now*—second persons come second. And truth? The truth of my labor lies less in the specific tale, adage, or idiom than in the way you work through these words, reader, these scraps of thought, to make what is real as real as real can be. Because there's nothing here that you don't already know.

Like I say, we've learned to laugh but *there's nothing funny about it. Nothing at all.*

My mother—my mother has a hard time laughing at any of this.

My mother doesn't like rats, or firecrackers, or horror films. Maybe she hurts too much. Of a German (Protestant) mother, Johanna Bentz, whose Chez Bentz features serviettes embroidered with an ornate B; of a Parisian (Catholic) father, Henri Alexandre Bourgoin, eventual Chef des Infirmiers for the potassium mines that dot the region around Mulhouse; my mother—a French citizen by birth, born in Landau, Germany, her communion held in the French Reformed Church of Wittelsheim, a silver Huguenot cross her keepsake (a matching cross of gold her older sister Ilse's)—my mother understands the territory perhaps too well, her Alsatian adolescence marred by loss of home and hope and refugee status in one's own country, her trek-to-be from one land of salt across an ocean of saltwater to a salt city.

She spent too much time in air raid shelters, detention camps. Too much time rationing food in the Alps, while the Nazis closed in on the "free zone." And too little time at the University of Grenoble, time interrupted by the war. As a teenager on summer break in the late thirties, visiting her girlfriend near Saarbrücken, she happens upon an SS uniform hanging in a family closet.

Out on the creek, late at night, busted moonbeams.

2.

Landscaping

> Surely it wasn't possible to be lost in the local?
> —Michael Joyce, "Storm Tossed"

SUNLIT DESOLATION, grey-white cloudbanks, empty with beauty. Far as the eye can see.

Mankind moves in mysterious ways. After the divorce, and after years of brand loyalty, brand names carry over. Head of household and two dependents wash their hair with Breck, lather-rinse-repeating till squeak. Are glad they use Dial. Rub their underarms with Secret, without embarrassment. Brush their teeth with Gleem, imagining ever-whiter teeth. Use Spic And Span and Mr. Clean on the floors, Tide on the laundry, Comet and Ajax on the pots and pans, Aunt Jemima on the French toast. Milk and orange juice in returnable bottles, from the dairy down the road. Eggs and potatoes a staple, garlic and onions always on hand.

Post-Briggs & Stratton, post-minibike craze, Mike and I ogle the ads and test drives in *Cycle Guide*. Jap bikes are in, Yamaha looking good. My father buys a shirt at Wells and Coverly, a pair of shoes at Nettleton. When he has the dough.

Up in the attic, jumble of assorted possessions. Remnants of Topper and Mattel and Hasbro call to mind Shoppers' Fair and Easy Bargain Center. We were spoiled as kids on birthdays and Christmastime, usually at discount prices. Family items mixed in—an old Viewfinder here, there an old Zippo.

Out back—where we park, and enter—runs the creek, banks covered with lush, rotting vegetation. Hold your breath, don't get wet.

A small shed protrudes from this rear-front of the house, over the entrance to the cellar. Walk down a few steps. Damp, musty. One light bulb, if you can find it. Support columns are tree trunks, debarked.

Another shed stands apart from the house, a garage of sorts. Full of junk.

Where we park, dirt and gravel, a yard worn clean to stone. Mud when it rains.

A hundred yards to the east and a bit north, the tiny house where the hermit lives.

And where Raphael Ave. dead-ends at the Thruway, the guy with the pigeons. The guy who threatens to kick the shit out of Mike if he pulls another wheelie down Raphael Ave. The guy my father ends up chasing across the field back to his house. His house always half finished—some new siding, part of a new roof. Next to it stands another, smaller structure, nearly complete. For his pigeons. They shit everywhere.

Everywhere in the neighborhood the sound of traffic moving past.

The entire scene a tiny wilderness, displaced. Just at city's limit, at suburb's edge. If this were the country, and it is, some would call it *white trash*. But we're not *that* white, so forget it.

Once in a while a hawk perches atop the large elm that hovers over the creek and reaches out to the house.

The TV is on upstairs. Always, or nearly so. At night and in the early morning hours a blue flicker fills the living room, my father's bedroom. A blue room full of smoke, and a yellow solitary ember. There are no curtains.

Rain patter on leaves, third day in a row. When I first open my eyes, the sound is familiar, relaxing. But this time another sound has been added to the chorus. I sit up and stretch over to the window, rolling my eyes down to see rain falling into water. PLOP PLOP PLOP PLOP. Water everywhere. I jump up, stagger out into the kitchen. PLOP. Mike gets up. My father is shaking his head, a cloud of smoke rising above.

We're flooded. PLOP PLOP.

Dirty lukewarm water out of our faucets. We still have gas and electric. My father has phoned the landlord, told him the gas furnace and hot water heater are underwater. The landlord has told him to phone the power company, and that he'll try to send some men over in the meantime.

Time to get out—the cellar might explode.

We dress up in shorts, the three of us. We walk downstairs—the creek has overflowed. The water line is six inches shy of the DeSoto's passenger-side window. We leave the car and wade up the street, the water brown and cool. A hundred yards up the street, dry land.

We use the apartment to sleep and to take a quick shower. We eat out. A few nights my father's mother cooks for us.

This goes on for a week before the waters begin to recede. No reports on the news. But you don't need the news to tell you that when it rains, it pours.

WNDR's ALL HIT MUSIC
week of July 10, 1972

1. Lean On Me .... Bill Withers
2. How Do You Do .... Mouth & MacNeal
3. Take It Easy .... Eagles
4. Brandy (You're A Fine Girl) .... Looking Glass
5. I Wanna Be Where You Are .... Michael Jackson
6. Happiest Girl In The Whole U.S.A. .... Donna Fargo
7. Conquistador/Salty Dog .... Procol Harum
8. If Loving You Is Wrong .... Luther Ingram
9. School's Out .... Alice Cooper
10. Too Late To Turn Back Now .... Cornelius Brothers

When there's no flood we eat over at my grandparents' maybe one or two Sundays a month. We eat early, around one in the afternoon. My gramma, who's been married since her early teens—she can barely write her name, but like so many of the older Italian women I've known, this woman can cook.

They say everybody can make a red sauce. Maybe so, but I've never had a red sauce like hers—dark red, oily, almost burnt. Same goes for her meatballs. A burp the next day is a burp you learn to appreciate. Sometimes it's chicken soup with tiny meatballs, or beef soup, an orange broth made with tomato puree and plenty of carrots, celery, onions, and elbow macaroni or thin spaghetti. A snow shower of grated Romano cheese to top it off. I love the broth-soaked onions best.

Sometimes my gramma cooks breaded pork chops and home fries. I eat five chops. Once in a great while my grampa broils us each a Porterhouse steak, with olive oil and lemon.

But usually it's macaroni with meatballs and sausage, on special occasions with pork hockies, bracioles, and chicken—a plate of meat stacked a foot high. I can eat two plates of macaroni, four meatballs, two sausage links, a braciole, one or two pork hockies, and of course bread—fresh Italian bread from Columbus Bakery, a place owned and run by Greeks. It's the best bread in town because they use brick ovens, the same ovens they've used for decades.

My grampa tells me to eat the tough, chewy skin on the pork hockies. He says, with my father translating, that it helps you live long. Mike likes the chicken.

Mike and I drink Coke, or orange soda (we say *soda*, an hour's drive to the west they say *pop*), and my gramma keeps urging us to eat long after we're full. We understand a bit of Italian. Sometimes we have a light salad after, and she motions to her mouth with her hands, speaking.

She's telling you it helps to clean the mouth, boys.

For me, maybe a hunk of raw *finocchio*—fennel. And then an anisette cookie or two. Mike and I call them *banana cookies*.

My father sometimes argues with my gramma while we're eating. He says she poisons him with her words, even as she feeds us. She complains, endlessly, about her health—her false teeth, her eyeglasses. Her digestion.

Ma, enough, OK? Ma—

My grampa shakes his head, shuffles over to the porch.

Mike and I turn on the TV. It's an old black & white tube set, the picture so-so, the image lazily distorting in a leftward motion along the top, rightward along the middle, and leftward along the bottom. Usually a western. Gramma and Grampa think the west still is this way. And their youngest son, my uncle Dominick, lives with his wife Dorothy and their two sons, Russ and Frank, in Cheyenne.

There's a war in Southeast Asia winding down, and a hurricane has hit the northeast, causing our flood. The DeSoto, the first car I'll drive alone, never quite recovers from the water.

On a summer day one year earlier, Brian walks over from the old neighborhood. Mike, Brian, and I decide to venture out, along the creek. Strange pungent odor, earthy and artificial all at once. We leap along the banks, pulling large rocks out of the oily mud and tossing them into the water. This creek, Ley Creek, feeds into the wastewater treatment plant a couple of miles downstream. Decades prior, my father's father helped pour the foundation for this plant. Years later I learn that the creek itself carries raw sewage and industrial solvents. This is where we play around, dirty our hands, soak our sneaks—shit creek.

Suddenly Brian yells. We stare into the water—a fish spawn. Brian grabs a stick and thrusts it into the water. It's so thick with carp he spikes one. We laugh. We spend the rest of the day poking around.

All three of us have diarrhea the next day, and for a few days after.

Dumbshits. And without a paddle.

But that first flood—we're selling fireworks that summer, Mike and I, for a little pocket money. My father buys the fireworks wholesale from his friend Norm Holstein. Mr. Holstein is his friend from their years together at General Electric, before the big layoffs in the late sixties, before my folks' divorce in '68.

Mr. Holstein is a "material handler"—he handles materials with his fat, calloused hands. When General Electric decides to move their huge Electronics Parkway operations to North Carolina, my father, angry and desperate, takes his severance pay, and Mr. Holstein takes the safe route—stays with General Electric at a wage cut. Mr. Holstein's way pays off, my father's way doesn't.

This is the same Mr. Holstein who, while sitting at our kitchen table at 112 South Dolores Terrace one morning, waiting to drive my father to work, once picked up a butter knife from our kitchen table and carved off all the icing on my mother's freshly baked kuchen. My father caught him at it.

Why you slob.

But he's a friendly guy, he likes Suzie, as many of my mother's American friends call her. And from a kid's point of view, Mr. Holstein is a big, jolly guy. In his later years he drops about fifty pounds, becomes a born-again Christian, drives even his religious wife crazy.

My father and Mr. Holstein don't see much of each other after the conversion. Mr. Holstein will outlive my father, but both men will die of cancer some twenty years later.

So on the fourth day of the first of two floods at 501 Raphael Ave., the three of us are getting set to wade up the street, where we've parked my uncle Sam's car—a beat-up '68 Buick he's letting us borrow.

He's an ex-con, my uncle. Three-time loser who did a twenty-one-year stretch in Attica and Auburn. Picked up for armed burglary, shot in the leg. It was entrapment actually, but who's keeping score?

The day he got out I spotted him walking down the street. It was the year I graduated from high school. He was wearing a toupee. My father didn't recognize him.

When we were kids, nobody told us we had an uncle named Sam. But I can recall my gramma putting together care packages containing food and such, and my father occasionally talking about driving to Auburn. And here and there dropping into Italian, *capisce?* I didn't make the connection till years later.

Before he dies, early into the next millennium, my uncle will have a leg amputated. To this day I wonder whether it's the same leg that took a bullet.

The three of us are going to carry along a bag of fireworks so Mike and I can try to make a few bucks over at the old neighborhood. Firecrackers, sparklers, and poppers—small candy-colored pellets that explode when thrown against a hard surface.

When we were kids, they'd make 'em big as jawbreakers—like this.

My father makes a one-inch circle with his thumb and forefinger.

And we'd run into the State Tower Building and smash 'em against the walls. POW! Helluva racket! The cops'd chase us down the street, but they couldn't catch us.

These days my father lights a firecracker by touching the fuse to the tip of his cigarette, and tossing it casually into the air. He chuckles a bit when he hears that POW!—brings out the kid in him.

We live in Syracuse by postal address, but in the North Syracuse School District, adjacent to the hamlet of Mattydale, which sits on the northern edge of Syracuse—on the edge of the city. But Mike is still attending Liverpool High, and I've just graduated from there.

Before my graduation, and each morning during the school year, my father drops us off at the bus stop at the top of Dolores Terrace. From there we take the bus to the high school five miles north—a growing suburb, Town of Clay, where my brother, decades later, will own a house. We don't want to switch schools, and friends. We do well in school, so school officials rarely ask any questions. And if they do, we play dumb.

Sometimes, when it's raining or cold, we huddle under the eaves of the large A-frame Presbyterian church, waiting for my father to pick us up after school. On weekends or days off, and whatever the weather, we walk over to Dolores Terrace to hang out with our friends. Once in a while my father can drop us off, leaving us with just the walk home.

We don't have jobs yet—we're not sure exactly how to get work, even how to get *to* work and back. We don't understand how work works. But we help out my father when he has work on the side—furniture to pick up and deliver. Sometimes he needs a hand on a house call, so one of us will tag along. Usually me. Most of my father's customers live in beautiful homes, and they tend to ignore me—which suits me just fine.

We pull on our shorts—we know the routine by now. My father is wearing shorts and rubber boots. This is a guy who's never owned a pair of blue jeans in his life, who works with stains and lacquers and sealers and sandpaper and putty knives wearing dress slacks and leather-sole shoes. Mike and I are making fun of him—he looks like he's had a bad day at camp. He's smiling and telling us to mind our own goddamn business.

We walk downstairs, pause a moment before getting wet.

C'mon boys.

We start to wade up the street. The water is brown and cool, like I say, waist high. My father has opted to carry the bag of fireworks under his right arm. We walk for a bit.

Dad, I think there's a ditch over there.

Topography is a bit of a problem at the moment, but Mike and I are both good at geometry. And my brother knows the terrain.

Dad, I think there's a ditch on that side, where you're walking.

Just mind your OWN GODDAMN BUSINESS—OK?—keep walking! PLOP.

One step, my father falls face forward—and down he goes, head and shoulders disappearing for a moment under the water's surface. His torso pops back above water, and he wipes his face off with his left paw. He's beet red, about to burst with every profanity known to humankind. The bag of fireworks is drenched. Mike and I are laughing so hard we're nearly pissing our soaked shorts. But even all wet, his is a commanding presence.

SHUT THE HELL UP, THE BOTH OF YOU! Now let's go—and make it snappy!

He tries to be angry. We keep walking, not quite holding it in.

Turns out the fireworks are OK. Mike and I make about twenty bucks that day—sell an entire brick of firecrackers to one group of kids in less than a minute. The waters recede four days later.

Though raised a Catholic, my father is hardly a religious man. Still, every now and then he likes to tell me that his name, like my name, is the name of a saint—St. Joseph, the patron saint of carpenters. During the Middle Ages in Sicily, when the island was suffering from an extreme drought and famine, the Sicilians prayed to St. Joseph for rain. They got rain.

St. Joseph's is the name of the hospital where my mother will draw her last breath.

Sunlit desolation, grey-white cloudbanks, empty with beauty. Far as the eye can see.

Front moving in, difficult to see through. Find a way through. A light grey, so light a grey approaching. The wind changes.

Branching every which way, branches. Redo rangey ell owgre en leaves. Water. Asphalt. Weeds. Earth.

Land of the Barge Canal, land of manufacturing.

The TV is a ritual—it draws us together, converting living space at Eastern Standard Time to information and entertainment. The three of us gather round to watch *The Wild, Wild West*, talking and arguing about this and that through the commercials, our talk spilling over into the show until somebody says SHHH, even if it's a repeat, or rerun. We can relate to James West and Artemis Gordon, relying as we do on one another, living as we do in the wild, wild east. The spoof spoofs us—and if things at 501 Raphael Ave. were just a bit more desperate, we'd be ripe for gimmicks, miracles. As it is, we're suckers for Kirk, Spock, and Bones.

Years prior, at 112 South Dolores Terrace, it was Elvis at the drive-in, and Connery as Bond. The drive-in is the ritual then—the Lakeshore, the Dewitt, the North, the Salina—and my father builds a small wooden bench to straddle the rear floorboard hump of our four-door baby-blue '52 Olds. The Olds with the Rocket V8, about which my father often boasts. My mother makes the backseat into a small bed for kids.

All in the Family: we laugh at Archie Bunker. Depending on where you're coming from, he's either a celebration of postwar whiteguyness, or a send-up of same. For my father, he's both. Mike and I lean toward send-up.

And movies, always movies, sometimes the same movies, over and over. And over. Especially the old black & whites. Oland. Weissmuller. Lorre. We identify them by the male leads—a Cagney film, a Bogart film, a Muni, a Boyer, a Cooper, a Gable, a Stewart, a Tracy, a Wayne, a Flynn, a Grant, a Peck, a Douglas, a Lancaster, a Mitchum, a Power, a Heston, a Newman, a Ladd. A few exceptions—a Karloff–Lugosi, a Hope–Crosby, an Astaire–Rogers, a Garbo, both Hepburns, a Fontaine, a Garland, a Loren.

My father says he was always told he looked like Dane Clark. And my mother, he always says she looks like Gene Tierney. Me, I imagine myself the director—calling shots, putting it all together.

My father—he has an annoying habit of getting up to take a piss right at the climax of a film. I figure it's because of his high blood pressure, driven higher by the suspense, even if he's watching for the tenth time one of the *Sons of Hercules* flicks.

We own a small BSA hi-fi, a gift from my mother on our second Christmas at 501 Raphael Ave.—AM/FM radio, turntable, 8-track. My brother and I play top-40s rock and pop when we're alone—once in a while I'll put on my Sinatra, Jolson, or Nat King Cole albums—and we begin buying 8-track tapes in addition to 45s and LPs. The Who, Herb Alpert and the Tijuana Brass, The Fifth Dimension.

My father likes Jerry Vale and Dionne Warwick. He especially likes the album *Promises, Promises*. The record sits on the turntable most of the time, gathering dust. When my father drinks, he plays it over and over, a new crackle or pop added with each passing season.

Downstairs live Freddie Holmes and his wife, and (eventually) two kids—one boy and one girl. Freddie drives a school bus for a living, can be heard below us every once in a while strumming his six-string and crooning c & w tunes off-key. He's OK, a bit of a sap. His wife is quiet in public, but

WNDR's ALL HIT MUSIC
week of March 27, 1972
ALL HIT ALBUMS

1	Harvest	Neil Young
2	America	America
3	Paul Simon	Paul Simon
4	The Concert For Bangladesh	George Harrison
5	Nilsson Schmilsson	Nilsson
6	Hendrix In The West	Jimi Hendrix
7	Baby I'm-A Want You	Bread
8	Eat A Peach	Allman Brothers
9	Fragile	Yes
10	American Pie	Don McLean

we can hear them screaming at each other every now and then—and they can hear us. And Freddie and his wife have a habit of not talking to their kids—they shout at them.

Once in a while Freddie mows the lawn. I think I might have myself, once or twice. It's a tiny lawn. Usually though we all just watch it grow. The weeds sometimes reach as high as the second-story porch. The house almost looks better this way, the blend of greens obscuring the chipped and faded wood siding. I suppose we appreciate the weeds, might on a given rainy day relate to them in some offhand way—we don't like it that we ourselves have been uprooted.

But the analogy to weeds can only inadequately capture our situation— we're not thriving here, and we're trying in our faltering human way to get out. Sooner or later the landlord, who owns the plumbing and heating supply company next door, sends over some workers to hack down the growth. And something feels right about this, finally—this exposure.

The inside layout: two adjoining bedrooms, a dining room, a small living room, a large kitchen, a bathroom right off the kitchen entrance, an attic. Living upstairs, we pretty much claim the attic for ourselves. The house has one nice construction detail—level tongue-in-groove wood floors. We find out a few months after moving in that it was Mr. Holstein's grandfather who built the place. My brother and I sleep in one bedroom, my father in the other.

After a few years, we decide that the living room is a waste of space. My father falls asleep on the couch most evenings—when he can sleep—leaving his bedroom empty most nights. And we don't enter the house from the living room anyway—the rear of the house has become the front, where we park, and we enter up the back stairs and into the kitchen. So we switch things around.

We put the dinner table—the same faux-marble-topped, yellow-vinyl-girded steel table we owned at 112 South Dolores Terrace, with the dent in it from the divorce—in the kitchen. We make the living room my father's bedroom. And we make the dining room the living room. This way my brother and I each get our own bedroom. My father still sleeps on the couch in the living room most nights, with the TV on. If he gets four hours of solid shuteye, that's a lot.

Summers Freddie's father and stepmother come up to visit from Florida.

Now Freddie's father is what you call a real pain in the ass. Loud, short, stocky guy—yells at his son relentlessly. He's outwardly friendly to us. His wife doesn't say much, doesn't seem to have much to say. I see his license plates, I call him Florida Orange.

Florida Orange visits towing a small camper. He parks his camper alongside the house, between the east-west fence and the storage yard, and runs an extension cord into Freddie's apartment for power. He and Freddie's stepmother live there, in the camper, for a month or two each summer they visit.

Florida Orange is nosy as hell. No matter what you're doing downstairs, FO has just got to walk over and ask you what you're up to. My father, though, is not a patient man, is an unsociable man most of the time, especially when he's impatient, or feeling put upon.

Hey—mind your OWN GODDAMN BUSINESS, OK CHAMP?

OK Joe.

That's the way it usually goes. For his part, FO yells at his son, the IDIOT, to do this or that whenever the two of them fiddle with their own cars, whatever.

One day my father is downstairs trying to fix the DeSoto radiator, which has sprung a leak. Now my father—he's an artist with wood, but as a car mechanic he leaves something to be desired.

It's not that he doesn't have a way with tools. When we're kids, he pulls out our baby teeth, when they get loose, with a pair of needle-nosed pliers, metal tips wrapped in his hanky. And he's gentle about it.

Still, as a car mechanic, he leaves something to be desired.

First, he's got to pull the DeSoto radiator to bring it to a garage, where he can have it brazed. After about an hour or so of struggling with the rusted nuts and bolts, he gets the radiator out. He's arranged to have his best friend, Dick Italia, drive him to the garage. The DeSoto used to be Dick's—he's given it to us.

Dick eventually decides, in his forties, to relocate with his family to Tucson, start a new life. My father thinks he's nuts. These days, I think Dick was ahead of his time.

Anyway, my father comes upstairs to tell Mike and me that he'll be back home in a while. We're doing our schoolwork.

He returns in an hour or so with the repaired radiator—cost him fifty bucks, a lot for us then. He's got the car positioned in front of the shed-garage that stands apart from the house. He's got the front end jacked-up, with a bumper jack, so he can get underneath it. He manages to reinstall the repaired radiator. After filling the system with coolant, he wants to check it for leaks.

Problem is that he can't seem to get the car started. He turns the key—nothing. He gets nervous—my father gets nervous easily when it comes to things mechanical, and he gets nervous easily these days. So he gets an idea—he'll place a rock on the gas pedal, and start the car by crossing the solenoid under the hood to bypass the ignition switch.

He gets out of the car and scrounges up a piece of wire. After stripping both ends carefully, he leans over the engine and crosses the solenoid. The car turns over, and the engine immediately fires up. Fires up hard—after all, there's a rock on the gas pedal.

What my father doesn't realize—doesn't until just *now*—is that the car is still in Drive—this is why it won't start. In a '60 DeSoto, it's easy to forget to press the Park pushbutton after you turn off the engine, and there's no safety that prevents you from removing the key while in Drive.

The car lurches forward, its back tires digging out. The bumper jack goes sailing, my father leaps out of the way. The car slams into the garage, knocking the entire structure completely out of square. The engine fan smashes into the newly repaired radiator, grinding its blades through the coolant passages. Now it's *really* leaking. My father manages to jump in and shut it off.

He walks upstairs, nearly out of breath, what hair he has left on his balding head pushed all over the place. He tries to tell us what's happened, but the moment Mike and I get one look at that exasperated expression, hair and all, we laugh ourselves silly. He cracks a smile—half furious, half laughing himself. That same smile that can turn wickedass mad in an instant.

But this time it's a sign of mischief accomplished. Like he was still a kid or something. Like back when he shoved the arrow up the ass of that bronze Indian statue that used to be a prominent fixture in that small park on the north side, near Court Street and Salina. The holes are still there in the early seventies, a decade before the statue itself disappears.

The landlord has the garage torn down within the year. Not a trace remains. From our point of view, this is a good thing. Now the lot is wide-open, room for more than two vehicles.

Double exposure.

OK, so life goes on—so what. What goes on within life?

What keeps us going?

Him?

If it weren't for you boys—

WNDR's ALL HIT MUSIC
week of July 28, 1971
TOP 6 LPS

1	Friends & Love	Chuck Mangione
2	Ram	Paul McCartney
3	Every Picture Tells A Story	Rod Stewart
4	Carpenters	Carpenters
5	Leon Russell & Shelter People	Leon Russell
6	Tapestry	Carole King

Sitting in any room, you can hear the occasional metallic PONG of empty gasoline storage tanks—the type they use underground, at gas stations—expanding and contracting with the weather. The storage yard is full of large tanks, gasoline pumps, fittings, equipment—it's like living next to a junkyard. And like all junk, this junk makes money—in this case, by helping to supply fuel. And in this case, for somebody else, always for somebody else.

When we move in, rent is seventy-five bucks a month, plus light and gas. When we leave, a dozen years later, it's up to one-hundred-fifteen a month, plus utilities. You get what you pay for. We get more, each of us, and less. More wisdom, more ignorance—whether poor or rich, some stakes remain the same.

One way in, one way out—you pull in, that's that. But once in, we're in for the long haul. In for the duration.

We'll be here just a few months, boys, till we get back on our feet.

Till we get back on our feet, water flows in, water flows out. But water seeks its own level. You stop the flow out, the level rises. Like tempers, like stacks of bills. You cut off the flow in, the level falls. Outages, want.

Over time it becomes as apparent as the first sign you see when you take the turn: NO OUTLET.

We're self-contained, the three of us, have no place to go but home. Still, borders may be crossed, just as emotions have been known to seep.

While we're still living at 112 South Dolores Terrace, waiting for our house to be sold, my father starts seeing Peachy. Her real name is Elizabeth, but she goes by Peachy. This is when my mother has moved into an apartment at Grant Village over in Eastwood, the same apartment complex she'll return to twenty years later, after moving back to Syracuse from Schenectady. The same apartment complex into which Mike and I will move my father, after my mother dies. Into my mother's very apartment in fact, the apartment in which he'll die a year and a day later, in my brother's arms, coughing up blood.

Peachy, who's been married twice—she seems all right. Mike and I are just teenagers. She's fun to be around. She's always teasing us. My father is around fifty, she's around thirty. Peachy and my father continue to see each other when we move into 501 Raphael Ave. She has a seven-year-old son, Billy, from her first marriage. Peachy and Billy, and their black terrier Grubby, constitute one border crossing.

Another is my mother.

My father drinks. He started hitting the bottle when the marriage started falling apart. On the surface of it, it's a simple enough matter—my mother wants a divorce because she's met another man. The other side of the story is another story, one that, at thirteen, I don't care to hear. Either way, it's hard to know whether my folks married for the wrong reasons, or for too few of the right ones. But one thing is certain: they share a history of struggle. My father met my mother in Europe, during the war. She was a French Red Cross worker, he a U.S. Army corporal.

The evening the divorce is finalized, I'm walking downtown with my father. He's brown bagging it even then. A light snow is falling.

Well that's it then, Joey.

Eyes welling up, jittery, slurring his words just a bit—I'll always detect his drinking through this slur—he tells me that my mother's lawyer, that sonofaBITCH, is disappointed that she's granted him custody of Mike and me. This has been our wish, but I wonder what my mother's decision means, will mean.

The man for whom she's left my father, Henry, is himself married. He's promised to get a divorce. (Never does. The last time I see Henry, he's drunk in a bar in Schenectady, wearing a leisure suit, hitting on the cocktail waitresses.)

(Oh Henry cut it out.)

(This is the year before my mother dies unexpectedly, of heart failure.)

* * *

Things with my mother are rough at first. I can't deal with her, and she can't deal with me.

Dad, come and pick us up.

But after our first year at 501 Raphael Ave., my mother begins to visit us monthly. She's moved to Schenectady, taken a job as main receptionist with General Electric CR & D—Corporate Research and Development.

Washing clothes at the Laundromat is not something my father, Mike, and I do well together. We wash what we can when we have a mind to—in the kitchen sink, with a washboard, or on the stove, in a large, white-enameled pot of boiling water. Otherwise, what gets dirty gets dirtier. So when my mother visits, the first thing she does is take our clothes and bedsheets and towels to the Laundromat.

The sheets on my sinking, creaky mattress are usually a yellow-brown by the time my mother arrives. My mother always shakes her head when she sees this, makes a maternal tsk-ing. But I keep my room pretty neat, the paperbacks on my metal bookcase, which I pick up at Economy Bookstore downtown, aligned squarely on the shelves, alphabetical by author and arranged by genre. Once every month or two I even dust a bit.

My father and mother are generally civil toward each other. Sometimes even friendly. Peachy usually makes it a point not to be around when my mother visits.

Aside from a few handfuls of friends from Dolores Terrace, and from school; a few relatives; a few instituted contacts, like teachers, the landlord, and the social service agency; aside from the mail, the stereo, the TV—always the TV—what carries conflict and care across the 501 Raphael Ave. threshold of these early years is the telephone. It hangs in the kitchen, on the wall next to the cupboards. We begin with a simple list, taped to the wall alongside the phone. Over time we scratch numbers on the wall itself. Since the wall is never painted, it becomes a record of names and numbers, of stresses and strains endured, of relationships made and broken.

My father uses the phone to regulate his relationships, his obligations. And Mike and I are usually caught in the middle.

I'll fix her ass. If it's Peachy, tell her I'm not home.

If it's Harris, tell him I'm not home.

If it's Beneficial Finance, tell 'em I'm not home.

If it's—

But Dad—

DO AS I SAY!

Landscaping

* * *

Sometimes, arguing with Onondaga County clerks and social workers, he loses his temper. My father is not a patient man, and he has zero patience for the ins and outs of bureaucracies. So he puts me, a kid, on the phone, and I argue with them. Some of them are OK, some not so OK. OK or not, it's not long before his Social Security number becomes a part of my active vocabulary. And it's not long before I, we forget select details of the aid we received.

How soon we forget?

But you want to know: did we receive public assistance? I'm pretty damn sure we did, I know we received food and food stamps. Medicaid. But technically—

Few official records exist to document, prove.

The proof is in the pudding?

Or in the puddles?

True, you can sometimes earn more working under the table.

True, these are good things to forget. Selective absorption, reflection, scattering.

Shielding, fabricating, Mike and I become his quick fix—an unhealthy resource to cope with his troubles, his addictions, we three a self-contained outpost. Him?—he understands only according to his own lights, he understands only that he works for us.

We're his sole source of pride, constant, a man with few friends, and fewer connections. Like everyone else, he knows how to get a hold of an illegal New York State vehicle inspection sticker. But rarely does he ask for favors, and rarely is he offered any. A few exceptions help turn persistence into luck, luck into persistence.

And later, when we can drive, when he's not at the Town Pump, my father spends more and more time on the couch. He's watching TV, he's not watching TV.

Hey, if you're going out, stop down at the store and pick me up a coupla packs of smokes and a six-pack.

But Dad—

Do as I say!

We do as he says. Because

If it weren't for you boys—

If it weren't for you boys—

If it weren't for you boys—

Sometimes we lie to one another. Sometimes it's the only way.

We learn to accentuate the positive, sure, but there's no way to eliminate the negative. No fucking way, not on your life—not there, not *here*.

In my mustard yellow high school yearbook, the *Hiawathan*, under my photo—smiling, hair parted to the side, horn-rimmed glasses, tie, jacket a bit large (it was my father's): *Engineering*.

I had no idea—it was my best guess, a way out. Or so I thought. My occupational placement test had placed me either as a bricklayer or an electrical engineer. I tell myself then what I'll learn part-time twenty years later, the hard way, no offense to bricklayers:

The hell with bricks.

And as for *Engineering*, it's just a best guess, another word.

What keeps us going?

WNDR's 40 HAPPENING HITS
week of April 24, 1970
THE BIG 10

1	Vehicle	Ides Of March
2	Reflections Of My Life	Marmalade
3	For The Love Of Him	Bobbi Martin
4	Make Me Smile	Chicago
5	Tennessee Birdwalk	Blanchard/Morgan
6	Miss America	Mark Lindsay
7	Little Green Bag	George Baker
8	Everything Is Beautiful	Ray Stevens
9	Let It Be	The Beatles
10	Woodstock	C S N & Y

A nightmare: we're in our old F-85, around 11 pm, on our way home. Mike is in the back seat, I'm sitting alongside my father, in the front. He's brown bagging it. The tie-rods are loose, the front end is shimmying and squeaking like mad. We're headed north on Interstate 81—we call it Route 81—after it leaves the city, on the bridge that crosses over Park Street and bends around to the northeast, toward North Syracuse.

My father is holding the steering wheel with his left hand, holding the bottle in his right. He's got one foot on the brake and one on the accelerator—so he can hit the brake *fast*—and he's stepping on it. As we hit the bend, the speedometer reads 95 mph and climbing. No seat belts. I'm pushing both my feet hard against the floorboards, leveraging my body back against the back of the bench seat, holding onto the window crank with my

right hand, my left hand clenched. Mike is sitting up, holding onto the back of the seat with both hands. A couple of empty Buds jingle under the seat.
Dad, slow down a bit.
SHUTUP!
A nightmare. With product placement—but weren't those Buds Black Labels?

My father picks up the F-85 after the accident on Route 11—U.S. Route 11—the accident occurring during formal divorce proceedings. My father is driving a shimmering turquoise 1966 Delta 88. The metallic green Olds wagon we'd owned for five years (with power rear window), the next in line after our baby-blue '52 Olds, had finally given up the ghost, and my folks had purchased the Delta 88 on the cusp of their marital problems. My father, you can tell, was an Oldsmobile man—just as we're a General Electric family—and the Delta 88 would be the last new car he would own.

My father is headed south, when a guy coming north swings a sharp left across our lane and into the J. M. Fields parking lot. It's true my father is a bit rattled about things prior to the crash, but still, no way could he miss the guy.

At the moment of impact, I can see the guy's cigar instantly crunched up into his face, just before our hood buckles. I'm sitting in the front, Mike is in the back. My father instinctively reaches his right arm over, holds me back against the seat. I can feel Mike bounce off the seat in back of me.

SonofaBITCH!

We're in the right, the cops agree. But the 88 is totaled. Enter the F-85—and it's beat.

Six months after the divorce, my father will get home late one night, drunk. Mike and I will learn that the F-85 has some front-end damage. The car will remain in the garage for a few furtive weeks, while my father and Dick Italia repair it. Mike and I will be instructed not to tell a soul about the damage.

Three years later, my father will walk into 501 Raphael Ave. a bit drunk, boasting about how, parked at the bowling alley, he backed up another in our long line of shitboxes only to discover that he'd backed *into* a guy who was parked right behind him, blocking him in.

I rammed that sonofaBITCHIN' assHOLE right out of the fuckin' way!
We'll all three of us laugh.
Ha-ha o-fifty-six ha-ha fourteen ha-ha nine-nine-seven-five. Hahahaha ha.
I struggle then, like now, with these stories. Just stories.

* * *

Sunlit desolation, grey-white cloudbanks, empty with beauty. Far as the eye can see.

Front moving in, difficult to see through. Find a way through. A light grey, so light a grey approaching. The wind changes.

Branching every which way, branches. Redorangeyellowgreen leaves. Water. Asphalt. Weeds. Earth. Social security?

Land of the Barge Canal, land of manufacturing.

A small step for one woman, landing, four falling apart
across the distances.

What keeps us going?

3.

Wicked Piss

Some come dark and strange like dying
— Joni Mitchell, "Songs to Aging Children Come"

STILL TOYING with the art of letters, of prepositional objects and subjects. Thought doubles back on itself, can simplify or amplify attribution and affiliation. The art of numbers looms, its truth tables and prose tabulations a risky proposition in these latitudes. A commonplace.

It's maybe twenty degrees out. We're all wearing CPOs, some are lined. No hat, no gloves. Sneaks, jeans, maybe a flannel shirt. The streets are icy, there's two feet of snow on the ground. Some drifts as high as four feet. Puddles frozen over, ice shattering as you walk across.

We're standing in Julie's garage, our hands in our pockets. A couple of us are smoking. The side door is open. It's five o'clock, getting dark. I've got to take a wicked piss.

The concrete floor is cold, damp. There's eight of us—Rick, Greg, Frank, Stan, Julie, Sally, Mike, and I. Sometimes Julie's brother, Rob, is there, sometimes her little brother Steve too. Once in a while Paul shows up, or Ben, or John. Much later, when weed becomes a constant variable, Jenny, Annie, Barb, Debbie, Sharon, a few other girls from nearby neighborhoods appear on the scene. And later still Lenny, Tim, Kevin, and Dan will hang out with us.

Julie and I have started making eyes at each other. I'm in my last few months of high school, about to turn seventeen. Julie is just finishing her freshman year. We're both virgins.

In our first moments alone with each other, two months prior, I'm wearing horn-rimmed glasses (usually busted and taped together), have longish unkempt light brown hair, make awkward motionings around girls my age. But not around Julie.

You're so handsome Joe.

You're beautiful, Julie.

She doesn't believe me. I don't quite believe myself. At seventeen I know only that she means everything to me. I peer into her violet eyes, stroke her blonde hair. She clutches me tight—too tight, I think.

I return the gesture, unknowing.

Some months later Julie will land her first part-time job, working the cash register at Ibsen's dairy after school. I'll stop by to talk with her in between customers. We'll talk about her family problems, and my family problems. We'll share secrets, and loyalties, and aspirations.

Now, strewn around us in cardboard boxes, are assorted animal parts. Over here, a box of deer heads. There, a pile of skins. The smell is what you would expect. A cat walks in and out. And a large dog meanders about. Julie's father is a taxidermist part-time, a line repairman days. He works out of his basement. He grew up on a farm to the north, hunting and trapping.

The Beckers' is the house on Dolores Terrace with the run-down vehicles. The half-completed addition. The beat-up boat out back, the rutted yard, the crumbling driveway. The four kids. The house at the end of the dead end, just a hundred feet from where Shane Drive intersects South Dolores Terrace to provide a right-angled way out of the neighborhood, to Buckley Road. The farmer's field borders one side of the house.

Years prior I run across, butterfly net in hand, warm wind blowing through my hair. My mother has stitched the netting to a coat hanger, attached the hanger to one of my father's long wooden dowels. But I find out the hard way that butterflies don't keep well in a jar.

Julie's mother hails from the south coast of England. She hates hunting and taxidermy, struggles with the mess of it, is always trying to keep up appearances. For some folks, keeping up appearances is the only way to keep up. A losing battle, but there it is. Me, I learned early on that style usually reveals a substance of one sort or another. I know little of fashion, and less of what it means to be fashionable. But it's clear to me that the unfashionable Beckers are loyal as the day is long, loyal to a fault.

Mrs. Becker is one of my mother's best friends. Like my mother, she met her husband overseas, while he was in the service. She calls us kids *little buggers*.

We're hanging out, shooting the shit. Sometimes, when Rick's parents aren't home, we hang out at his place. But the Beckers' is the one place where we can hang out when Mr. and Mrs. Becker *are* home. The one place where *our* appearance, as a group, is appreciated, even as our appearances go without comment. Long hair, soiled jeans, cigarettes, whatever. The Beckers are tolerant people, and extremely kind to animals—the animals they don't eat or stuff. Appearances to the contrary, it's a consistent way of life.

I've gotta take a wicked piss.
Greg disappears around the side of the house.
Oh bullshit.
Bullshit my ass.
Have you heard the new Zeppelin?
Yeah.
Sucks moose-cock.
Fuck you it does.
Yeah, it does—sucks moose-cock. One is better.
Frank! Geez.
Sorry Julie.
Delmedico, you've got to get your head examined.
Fuck you Joe.
Fuckin' Delmedico.
Fuck you Frank.
Rick, don't listen to those guys.
I know Julie—they envy me.

Rick is right, in a way. He's got a bad case of acne, but he's the most outgoing, the most willing to experiment socially, the most considerate, and the most willing to change. And because he's willing to change, he'll stay close friends with Mike and I, over the years, as we change. Change, and remain the same.

Rick's father is an electrician. Rick will think about this career for himself, but end up a psychologist. And he *will* specialize in getting heads examined—head trauma cases on a postdoc at Berkeley.

This what you call *destiny*?

I've gotta take a wicked piss.
I walk around the side of the house. Piss out my given name, in cursive. I wonder if Greg has. No signs.
Aw fuck it.
He can't drive that goat for shit. Remember when he lost it and ran into that telephone pole down the street! On his fucking permit, that numbnuts.
Dumbfuck.
Shit, fuck this noise. I'm hungry.
You guys are impossible!

Sorry Sally.
Just then the kitchen door opens—it's Julie's mother.
Julie, we're going to be eating soon.
OK Mother.

It's getting dark.

Dinnertime, and time for Mike and I to head home.
See you two douchebags.
See ya later.
See ya. Hey wait a minute—you guys wanna go get some shitfood with us tomorrow night? I mean, it's Friday.
Where?
Carroll's. I could use a clubburger.
McDonald's?
Fuck that—Red Barn—they have chicken.
Naw, Burger King.
OK.
OK. Why don't you meet Mike and I there?

You want good grammar, or bad taste? *Meet Mike and I,* improper, an uppercase symmetry to sound a prevailing construction. And in our case, *here,* it's the only construction that'll do. The subject was always Mike and I, inseparable. For now.

OK.
Later.
See ya.
See you tomorrow, Julie.
Bye Joe. Bye Mike.
See ya Julie.
The crowd breaks up. Mike and I begin our walk.
Joe, wait up—gotta take a wicked piss.
I wonder.
We cut through the backyards where we can. Over to Orchard West, and down the street to the end. Eyes out for the Orchard hardasses, who'll cobble your skull just for kicks. Between two yards to the new access road. Down the road and, if present, across the slat over the drainage creek. If not, we look for a place to cross. Through the hole cut in the wire fence, and across Northern Lights Circle. Up the hill in front of K-Mart, and down Route 11 through Mattydale another mile. Past the Hollywood Theater, and down the grade. Up

over the Thruway bridge on Lemoyne Ave., down the embankment, and we're home. Two-and-a-half miles—maybe a half hour in this weather, at this pace.

I'm freezin' my fuckinassoff.

I'm freezin' my fuckinutsoff.

That summer before college, my last wageless summer, is when my mother buys me the Camaro. When college starts in the fall, it takes us some time, my father, Mike, and I, to work out our schedules. When we do, I'll walk down the hill from Syracuse University, maybe a mile to Erie Boulevard. It's around three o'clock. If I have the fifty cents, I'll take the bus to Hanson Interiors three miles down the road, or hitchhike if it's running late and I'm cold. I'll get the keys to the Camaro from my father, then drive twelve miles to Dolores Terrace to pick up Mike from high school. Mike and I will get home around half-past four. Around a quarter to six, I'll drive the seven miles back to Hanson's to pick up my father. Then home.

Between picking one another up, and errands, and just plain driving around, the odometer jumps about a hundred miles a day.

The Camaro has G-60s on the back—the popular new Wide-Ovals. We don't have the money for snow tires. No money at all to keep the car up. I should have asked my mother for a sedan.

Instead, we slip and slide all winter long.

Some midwinter nights we head over to the huge General Electric parking lot off Electronics Parkway, now two-thirds empty due to layoffs. Mike's the first to get this idea into his head. At 30 mph, he'll flip the Camaro's steering wheel one way or the other, sending the car into a mad spin across the ice-covered asphalt.

We blow a lifter. My first real attempt at fixing a car. Takes me eight hours of fucking around. You have to pull the intake. A bit of a job, especially outside, in the snow. We have a fire going in the garbage can, to keep our hands warm. We're out back—the rear-front of 501 Raphael Ave. I don't have all the right tools. But Frank helps out—he knows a lot.

I have to take a wicked piss.

I walk around the side of the house.

I end up getting it right, except for that bastard gasket near the firewall. A week later, an oil leak tells me it's slipped out of place. Have to do the entire job all over again.

After a while, I'm the house mechanic. '69 Camaro. '68 Chevelle. '67 Bel Air. '71 Impala. '74 Nova. '68 Caprice. '68 Firebird. '71 Cutlass. '66 Falcon. Mostly Chevys, but whatever rolls in, rolls out.

When it comes to mechanical work, rust and wear are par for the course. Everything fits just so, but few things come off or go on without some effort, even though the better mechanics boast the fewest bruised knuckles. I'll start with sockets and wrenches, often end up using my favorite hand tools—a pair of Vise-Grips, a pair of Channellocks, and a flat-head screwdriver, with a fine edge. And of course, my 3/8"drive Husky ratchet and socket set. Helps to have air pressure around, just in case—an impact wrench can work wonders.

Starters. Water pumps. Radiators. Belts. Batteries. Motor mounts. Brakes. Shocks. Ball joints. Carburetors. Valve jobs. Tune-ups—plugs, timing, points, cap, rotor, wires, and I check everything else. (In the thirties, you got those spark coils humming "in tune." In 2001, it's all r & r, modular, and less pollution. Long as you have a ten- or twenty-thousand-dollar computer for troubleshooting, you're all set.) Oil changes. (I collect used oil in a pan, dump it in the weeds out back. No harm done.)

Bodywork. Not just my car—Mike's/Julie's/Peachy's/Helen's/whoever's. I do it myself. Or we do it ourselves—usually a helping hand around. Not always though.

Once in a while I'll use the do-it-yourself garage over on Route 11, near Switz's. The guy who runs the place, Rich, is a good mechanic, and he's usually on hand to help out. But he shuts down after a year or so, just a few years before self-service gas stations begin to take over.

Eventually, when I'm in a jam, I'll use Frank's gas station-garage on Old Liverpool Road. Really his brother Tony's gas station. Really their father's gas station. It sits right in front of the Poor House North, one of our favorite bars when we turn eighteen. They're strict about proofing you for ID.

Tony I've known since kindergarten. I can recall at five years old giving him all of the baseball cards I'd collected up to that time. Why, I don't know. Tony is a real brawler—bites the back of his wrist when he gets mad—and a real mechanic, owns a full set of Snap-on hand tools. He and Frank will spend years getting their airframe-power plant licenses, then their pilot's licenses. Their family is well-off. Their mother is a warm, hard-working woman from Italy, their father, a down-to-earth guy and an aircraft mechanic himself. He dies in his late fifties, from respiratory complications.

Frank is still flying to this day, Tony is back to motors.

Sometimes I get caught, end up lying on my back in a parking lot doing car repairs. Happens once in the pouring rain at K-Mart. I take the starter off, bring it over to the K-Mart mechanic. Fat fuck, kinda goofy. But knows cars. Tells me to have a look at his '69 Camaro. It's a bit of a wreck, smashed in here and there, but licensed. No markers on the outside to indicate engine size, nothing. Fat tires.

I pop the hood. Fucker's dropped a rat motor into that small engine compartment. What you call a tight fit.

How'd you get that big block in there? Some of the '69s came with 396s, I know—

Yep, that's it. But it still took a little doin', this'n that.

I bet.

I lose control every now and then, but it screams.

Yeah.

The first summer after college, I decide to fix up the Camaro. Lose the rust, repaint it. The pale yellow is nice, but not mean enough. I choose a deep green, with subtle gold flake—DuPont acrylic lacquer #5489LH.

My father, Mike, and I spend days filling, sanding, prepping, and painting. Dirt all around us. Drop anything, you have to blow it off with compressed air.

The compressor is located in the shed over the cellar entrance. The same compressor and water (now air) storage tank my father used in the garage at 112 South Dolores Terrace, for furniture. He got it from Dick Italia.

When we move into 501 Raphael Ave., Mike and I busy ourselves with cleaning out this small area. We dig out and level the dank soil, rotting paper, assorted junk. The ground full of holes. Rodents.

And we set up the compressor. A loud, ancient single-cylinder. I drain and replace the oil, clean up the motor brushes, adjust the belt, and it hums. Eighty pounds air pressure max, and the pop-off blows.

Now, my father uses this same compressor outfit, with his Binks 29, to lay on gleaming coat after coat of 5489LH on my—our—Camaro. In between coats he whistles "Swinging on a Star" and "If I Only Had a Match."

When we finish wet-sanding and rubbing it out, it sparkles. Mike masks the pin-striping. Exact.

And it's getting dark again.

It's not that fast a car, it's not quite what you'd call tricked-out, but it moves along OK and looks pretty damned good. 307 cubic inch V-8, the low performance motor Chevy puts out that gets twenty miles per gallon if you take it easy. Ballsy three-speed Turbo-Hydramatic tranny. Fun car, seats four comfortably. White bucket seats that I bleach clean every week. Blow out the carpeting with compressed air. Clean the outside so thoroughly that I can run my hanky over it with little more than a smudge on the linen. This is the stuff that makes men out of boys, boys out of men.

I take the curve on Buckley Road, just before the Taft intersection, at 85 mph, with seven friends crammed in. 55 mph speed limit then, 35 mph these days.

When the light turns green, I hum past the dairy, get it up to 80 again before Buckley Road Elementary, what they're now calling Nate Perry Elementary, after our first principal. A five-minute walk from 112 South Dolores Terrace.

Perry was a fair guy, a tough guy. Kept a large wooden paddle in his office.

While we're busy taking care of the Camaro, we've got a problem upstairs. The bathtub walls, especially around the plumbing fixtures, are rotting out badly. This is not something I know how to fix. And my father is not a carpenter, has no experience with drywall.

But more importantly, it's not our place. And when it's not your place, it's your place only to spruce up, not to fix. Especially not when you're living in a shambles.

Some folks, I know, feel differently about this. Even poor folks. But in our case, as a rule, we neither spruce up *nor* fix. Chock it up to our collective despair, or just call us bad housekeepers.

We start to paint the apartment, get halfway through the old living room, now bedroom, soon-to-be game room—and stop. We lack motivation. We believe that, worst case, we'll be living here for another year, and that's it.

And another year, and that's it.

And another year, and that's it.

Another lesson.

My father tells the landlord—Harris—about the bathroom situation, but for a solid year Harris refuses to do anything about it. John Harris is not a bad guy—lets us fall behind as much as six months in rent—but he's only a mouthpiece for the man in charge, whom we never do meet. Assuming there's a man in charge.

And now water has begun to drip downstairs, whenever we take a shower. Right on top of Freddie's kitchen table. Freddie calls the Board of Health.

Workers appear within the week, tear out the old shower stall walls and install new mildew-resistant material all around, caulking the seams.

This is about the time Mike is working at Tony's gas station, pumping gas, and Tony is losing money, lifting it as he does from the till—summer of the so-called oil embargo, lines at the pumps without end.

Three years later the new walls begin to rot out.

Even when you have to take a wicked piss, it's tough not to notice the walls.

By the time we repaint the Camaro, I'm spending most of my time with Julie. Evenings I sit on the couch at her place, and watch TV with her and her family while doing my schoolwork. Her parents treat me well—feed me, pamper me even. When her parents and brothers go to bed, Julie and I develop our own, more profound version of statics and dynamics.

Our relationship, so many relationships, snippets of pop and rock lyrics.
Hello, it's—
Thinking 'bout the times you—
We could stay inside and play—
But there never seems to be enough—
I know a little 'bout—
I didn't mean to—
In and around the—
Hot fun in the—
You make the—
I really should be back at—
Can't ya—
Lookin' for—
Don't fear the—
A little bit of—
Workin' on mysteries with—

I land my first summer job, at Carrier Corporation. My first real job, with a paycheck. I apply, I'm interviewed, I get the job—just like that, without knowing anybody. Almost too good to be true.

I work in an office area along the perimeter of Carrier's large warehouse. My job is to help create a new inventory system for industrial air-conditioner parts—paperwork. They pay me a decent salary, and I befriend a guy there named Mark. Mark is a thirty-year-old, and into cars. I tell him my driveshaft universal joints are shot. He tells me to drive out to his place, in Chittenango—Shitandthengo has bragging rights as the birthplace of Midwest writer L. Frank Baum—and he'll give me a hand replacing them. I do, and he does.

Beautiful little falls just outside of town, one of our favorite haunts.

Late that night, driving back home down Route 5 around 70 mph, it's pitch black. As the road dips, the headlights of cars coming at me form a ceiling of light over my head.

*   *   *

Like the rest of the men and women with whom I work, Mark doesn't seem especially thrilled about his job, though he's grateful for the pay-check. Some days he shows up half in the bag. But in the bag or not, he's a good worker—when he's working. Most of us manage to bullshit away a couple of hours a day at least—it's one way to cope with the tedium of this work.

At least, this is what I learn on my first real job. To me, everybody looks older than his or her years, and frustrated. If you ask them whether they'd rather be working or sitting in the sun, they'll look at you cross-eyed and tell you to fuck off. But if you ask them whether they'd rather be working or unemployed, they're all happy as pigs in shit.

Once in a while I meet Greg for lunch. His father has landed him a summer job at Chrysler, working on the factory floor. His father has worked there for two dozen years, and will die a quarter century later, after forty years in the shop. This is not long after Greg's mother, also a factory worker, dies, and Greg will spend a lot of time sitting in the garage of their house—now his and his brother's house—smoking, drinking, waiting.

While we're both working around the corner from each other, Greg and I meet at the Friendly's Ice Cream down the road in Eastwood, or at Doug's Workin' Men's Tavern on Thompson Road, a place that sells real horsemeat sandwiches. Greg wants to up and join the Navy after high school, like his brother. And like his brother, he will. But unlike his brother, he'll enroll in college after—Le Moyne College, just a few minutes' drive down West Genesee Street from Syracuse University.

Greg and I spend one wild evening that summer trying to rescue Julie and her friends Debbie and Sharon. Rescue, that is, to our eighteen-year-old, can-drink-and-will-drive way of thinking. The three girls have coordinated things so that their parents think they'll be staying overnight at another girl's house. Truth is, they're headed up to Salmon River Falls, fifty miles north, to camp out and party.

But Greg and I get wind of Julie's folks getting wind of the plan. So it's up to us to hustle our asses up there, find them—somehow—and bring them back.

I'm driving the Camaro. We head north on Route 81, and as my speedometer climbs above 100 mph, Greg buckles his seat belt.

Nightfall.

Foggy in patches. We fly through some lane work and construction signs at 110 mph, make that first forty miles in twenty-five minutes. After the exit, we're stuck on dark country roads, and I'm hauling ass, doing 60 and 70 around hill and dale. Pretty hairy.

We finally locate the falls area—it's not clearly marked, it's not an official campground, the land is owned by the power company. I drive down the dirt and gravel road, and park. Greg and I stumble around, smell of pot strong in the air, everyone toking up around campfires that flicker against the trees and brush. You can just hear the river off in the distance.

Can't see a fuckin' thing.

Neither can I.

Did I ever tell you the first time I had a toke?

Uh-uh.

Me and Julie drove up to Westcott Reservoir one night. Campfires dotting the hill, kinda like this—end of the counterculture.

Greg laughs.

Anyway, we walk up to this one fire—and check it out—it's my father's friend's son—Mr. Holstein's son, Charlie. He offers me a J. I take one toke—my eyes watered. Harsh.

Didja get stoned?

Yeah, right.

Greg laughs, his laugh mingling with the voices in the darkness. Miraculously we locate Julie and her friends. Takes them five minutes to get their shit together, and then the five of us head back to 81 and south at a reduced speed of 90 mph. I pass three cops in the median at around 85, but—another miracle—not one nabs me with their radar. It's like we're invisible.

Our rescue works, and the parents, all of them, are left scratching their heads—suspicions aroused, but that's about it.

A year later Debbie will become a Jesus freak—that's what we call them—and invite us to a Christian rock concert. Not too bad, but those fuckin' lyrics, man.

Early evening, two-and-a-half years after I first spot the Camaro in the classifieds. Mike is driving, I'm sitting passenger side. We take the side streets up to Route 11, pop out near Switz's. We're trying to take a left from the side street, crossing the two lanes. It's two lanes coming at us from the left, one from the right. The light to the right turns red, and a pickup on our left pulls up short of the side street, leaving room enough for us to drive through. The driver waves us on. Mike looks to the right—no cars coming.

Just as he pulls out, turning left around the pickup, a large car comes blurring down that outside lane.

No Mike wait!

From my angle, I can just catch the Caddy coming, but I'm too late to do any good. The Caddy hits us solid, front left fender. A loud thud, the sound of metal folding, and we're jolted to a stop.

Our knees bruised, no real injuries. Fifteen minutes later, the cops arrive.

They say we pulled out, so we're wrong. We say the guy was speeding. But just try to prove it. I look at Mike.

Accident experts my ass.

But OK—we pulled out, we put ourselves at his mercy. Something our old man tells us never to do.

The Camaro has to be towed. Frame bent a bit, front end twisted. Totaled.

When Mike and I get home, my father nervously searches through the six-inch stack of bills on the coffee table. The insurance notice is buried under the pile. Our policy has just expired.

We all have to take a wickedass piss.

What's new? Broke, you take your chances, go for broke.

The Camaro is insured—or not—in my father's name. So he loses his driver's license for six months, Mike gets a formal warning, we're liable for $1,800 damage to the Caddy.

You're crazy if you think this will stop my old man from driving. Or Mike.

Within the month we end up finding the first in our long line of shitboxes. Funny, but I can't recall the make.

And you're crazy if you think the guy with the Caddy will collect on his judgment anytime soon. Or will collect anything close to $1,800.

Good luck mister.

It's around this time that Mike and a few of our friends begin regular break-ins at rock concerts in the War Memorial.

I try it once. We wrench open the southwest door, a struggle ensues, and I step in over a sprawling mass of bodies. I try to remain calm, walk slowly, end up walking right into a charging throng of cops and concert security personnel.

You see anyone breaking in here!—

I shrug my shoulders, shake my head.

No.

And I'm in, just like that. I find Mike a few minutes later.

You run into that horde of pigs?

Yeah, but I bluffed 'em.

Fuckers. Why don't they go stop some real crime.
Yeah, fuck it. Let's catch some Canned Heat.
We're both chuckling.

It's also around this time that we take a drive, a bunch of us, up to Lake George. It's a bit of a hike, from an east coast perspective. We plan to spend the night in a campground.

Me, I'm not much for camping out. But Mike is really getting into it. Jackie and Jenny and Rick and Mike and Julie and me and I forget who else. Mike is seeing Jenny. Jackie is her typical pain-in-the-ass self. Brings a squirt gun along, won't let anybody get any rest. I end up throwing a bucket of water over her, accidentally dousing everyone in the tent. Not exactly a wise move, OK.

OK—we're all young. All of us. All of us on the verge of growing up for good.

And it's two years later when we spot the Camaro. *In niggertown.*
And it's getting black out now. New moon.
And I've been putting this off. I knew I was headed here, but I've beat around the bush some, taken those back roads of which some of us are so fond. Cars and chicks—a certain girl, really. Love in the back seat. Clichés I've lived by.

It begins at dusk, with my father, Mike, and I driving through the south side, taking the long way home. Some old Chevy, might have been one of the Chevelles. And like other north-side Italian-Americans and north-siders, my father calls this part of town *niggertown.*

Niggertown, boys.

He shakes his head in disgust.

As a kid, this word for me is not a judgment, not a controversy. It's just a word designating a place—the south side, a part of downtown south of where my father grew up, south of where my grandparents still live. Where niggers live.

Nigger. As a kid, I don't quite know what the word signifies, but I know who it signifies. Not me, not my family, not my friends, not my friends' friends. End of story.

Eenie meenie miny mo / catch a nigger by his toe.

Wanna niggertoe, Joey?

He hands me a Brazil nut.

South of the south side, in Nedrow, one of my favorite spots, the Salina Drive-In. Has a playground in front of the screen. South of Nedrow, the

Onondaga Indian Reservation, 7,300 acres five miles south of Syracuse. The Onondaga, or Haudenosaunee—People of the Longhouse, Keepers of the Central Fire, for whom descent is matrilineal.

Maybe this is just a tribal yearning on my part—too many Wayne westerns as a kid—but what horse sense I can lay claim to tells me that the closest you'll come to the Iroquois lay of the land is heading north toward the city up Route 81. Syracuse will appear around a bend, cupped between hills, a recession dotted with a couple of twenty-story buildings at the northernmost edge of which is Lake Onondaga.

Or as we call it, Onondaga Lake. Today, winding through the city on 81, the white mushroom Carrier Dome of Syracuse University will loom to your right, where old Archbold Stadium used to stand. And if you continue north on 81 toward the town of Salina and the village of Liverpool, and as you drive over the bridge that rises above the farmers market on Park Street, what now dominates the view is the new mall they've built on the south shore of the lake—atop where the old junkyard used to be, alongside the wastewater treatment plant, and conveniently located across the street from the even newer P & C Stadium and consolidated bus/train station. As one might have imagined, convenience is a key word circa 2003.

I know, you're anxious to get going. So am I.

My father has lamented the demise of downtown for years. When he was a kid, he says, the north side was more *colorful*. More small shops, fewer dives—poor, but vitally so. The center of downtown, Clinton Square—the site today of a new outdoor ice-skating rink—was only a few blocks away from Loew's, the Paramount, the RKO, the Eckel, the Rivoli (the latter before my time). Woolworth's was a true five & dime, not simply a bus stop, and the Busy Bee was a hive of activity. And there was Edwards' Toy Store, connected by a tunnel under Clinton Street to the main department store.

Even during *my* childhood, walking through this tunnel and up into the toy store seemed a kid's paradise. I loved downtown during winters especially, a light snow hitting my fresh face, alive to the city, to the possibilities of the world. As my father describes it, downtown was a place for kids. Tough kids, but kids. But as both he and I understand it, *white* kids.

By the late sixties, after completion of Interstate 81 and construction of Route 690, the new overpass creating an intangible zone separating North from South Salina Street, downtown had become a place of businesses on the verge of bankruptcy. Of sixties-style architectural facades. Of ugly parking garages. Warren Street becomes the prostitutes' strip, and the Chimes Building designates downtown's south border—if you're white and middle class. Another block and there's a triple X-rated movie house, the Civic Follies. Another block, a social services center. Another, and it's suddenly the south side. Like magic.

Just about there, bear with me a moment longer.

A decade later, the city will begin to recuperate and revitalize the aged, red brick buildings and streets, eventually reinstalling the old forties-style street lamps in the Armory Square area. It might be too late. For many of the old mansions that once lined James Street, it is too late. For most of the elms that once lined the streets of the city, it is too late. For those of us who grew up trying to shake the feeling that all the foundations had been poured, it will always be and it will never be too late.

But for now—for now my father directs his anger at the people he sees. Lost on him are the social consequences of "urban renewal," of the gasping American dream. Lost on *him*, a victim of such desperate hope, coughing and sputtering even then, even then near that two-decade-long end of his modest rope.

But after all, this is the guy who walked out of *Midnight Cowboy*, calling it "filth." This is the guy who celebrates *his* father's vision of the city, back when the canal still ran down the center of Clinton Square. This is the guy who refuses to let go of the world he knew as a kid—the crazy world that sent him overseas as a young man, to fight a war. Like all worlds, a fractured world, a world that teaches you that there's no place like home, even when home, however vital, is chock full of requisite fractures, comfortable exclusions, desperate voices.

Now, says he, there's only creeps and cripples and fags.

Creeps and cripples and fags—and *niggers*.

The south side is a part of town we steer clear of, drive through only when necessary. Or to gape. Down here is where we used to eat fish on Fridays, at McCarthy's. Down here is where we still eat fish once in a while, at Bill's Fish Fry, if we don't feel like going to Jim's. McCarthy's closes down around the time Lee Alexander becomes mayor.

Twenty-five years later, Alexander dies in prison.

Down here, this part of town—Mike and I once pumped ourselves up to walk through this part of town. We dared ourselves, imagining it a grand adventure. I'm having a problem with wet dreams at the time, with soiling my underwear. My father bitches me out but good. Peachy tells him to leave me alone.

Mike and I walk through without event—no one bothers us. Just a few stares. Or so we imagine. Some years later Mike and Rick will walk through the south side distributing promotional materials for McGovern.

Here we are, in the dark once again.

Julie and I are parked at the dead end of a deserted suburban street not a mile from her house. It's a new moon.
You sure this is OK?
I'm sure.
We climb into the back seat, the streetlight behind and above throwing our shadows across the interior. We're careful with each other, a bit awkward, helping with each item of clothing. The vinyl is cool, soft. Each glancing touch of warmth brings a tremble.

We indulge ourselves, nervously, our bodies still new this close, in this newly foreign space. Our contact means something, means everything, is beyond us.

Time to head home.

And the three of us are driving through, cutting through. Mike and I are a bit nervous, sense that we don't belong, try not to stare. Not my father. He understands that staring is a form of entitlement, yet he stares. Gawks—defiantly. He just doesn't give a shit, but I'm wishing we were invisible.

At a traffic light, three black kids in their teens cross the street in front of us.
Wanna see me make some peanut butter?
He guns the engine. The teens turn and stare at us, but keep walking.
We, all three of us, laugh. It's an old joke by now, and besides—for Mike and I, laughing relieves some tension. Not my father. My father is brown bagging cheap red wine, one hand on the wheel.
Hey, there's the Camaro!
Mike has spotted it, in a gas station lot. I turn and look at it. It's now a primer gray, with helper springs in the rear, and an inexpensive brand of custom rims all the way around. The word comes to me as if by instinct.
Looks like shit—they *niggered* it.
A mile down the road my father stops at a gas station. He has to take a wicked piss.

4.

Games People Play

"I wonder how many of you know," she began, "that we are in the
Long House, the ancient domain of the Five Nations of the Iroquois."
 —Tobias Wolff, "In the Garden of the North American Martyrs"

AFTER THE DIVORCE, my father takes Mike and me to shoot pool, usually on weekends. My old man can shoot a good stick, especially eight ball. Runs eleven racks one time. He's always out of practice, can always kick our ass. First place we're regulars at is on South Bay Road, up in North Syracuse. Brand new full-size Brunswick tables, half-inch slate, green cloth. Real nice. But it shuts down in '69, like so much else in that region.

Second place is MoVito's, a pool hall on Old Liverpool Road. Not so nice, but decent sandwiches.

We read: not New England. Europeans flock to this place with the growing antebellum salt industry, an industry facilitated by the new Erie Canal. There are no blue bloods in this neck of the woods. These are boat people, refugees, wanderers—another kind of pilgrim.

We read: Onondaga County, nineteenth century: revivalist fervors, flare-ups of intense faith and conviction.

Finger Lakes: "the burnt- [or burned-] over district."

Post-Iroquois New Englanders: progressive and industrious, strident and iconoclastic.

Seneca Falls: Elizabeth Cady Stanton, first Women's Rights Convention, 1848.

Homer, Dresden: birthplace of Amelia Bloomer, "bloomers," first U.S. newspaper designed for women; birthplace of attorney general of Illinois and orator Robert Green Ingersoll, "the Great Agnostic."

Auburn: Harriet Tubman, abolitionist, a chief conductor of the Underground Railroad. About a century before my Uncle Sam did time there, I read.

Oneida County: revivalist Charles G. Finney. His convert, John Humphrey Noyes: the Oneida community. The Millerite movement, various Adventist sects.

Seneca, Wayne counties: the Fox sisters; Joseph Smith, founder of the Mormon Church.

Further to the west: ex-Mason William Morgan, murdered? The Anti-Masonic political party.

To the east, in Herkimer: cheddar! And along the Hudson Valley: the Shakers.

We read, reckon: Syracuse, the hub of this scarred region, is located approximately at the center of Central New York. We local readers, stigmatized whether we like it or no, call it the asshole of the state, and believe it to be a good place for assholes like us.

Traditionally, upstate New York is dairy-land Republican, and downstate, the Big Apple, urban Democrat. Syracuse, with its large union labor force, can swing both ways. Either way, it's what my buddy Frank's father likes to call a *big hick town*, the surrounding rural communities—gently sloping hills, farm fields, and woods at the outskirts—providing the conventional and stubborn myth of white Americana that tends to obscure the reality of the region's inhabitants, especially the sheer variety of immigrant-ethnic populations peppering the city. You have the Italian (used to be one out of every seven residents), the Polish, the German, the Jewish, the blacks, and the Lafayette Apple Festival twenty miles to the south. These days the Vietnamese have displaced some of the Italians, and some of the Italian poverty has changed hands.

The city's regional theater, its more cosmopolitan elements—epitomized by a few colleges, especially the university up on the hill, and the same old same old PR campaign to celebrate the city's diverse strengths—these have never entirely offset the working-class bedrock of the area. Whatever residue of snob appeal or progressive social thinking you can lay your hands on, this is a place where it's less important to buckle up than to buck up, a place where there's always incidents between campus students, many from downstate, and townies. It's a tough place, all told, a rough place, only a fifteen-minute drive in any direction to your favorite redneck bar, with jacked-up pickups out front and guys with baseball caps inside. Depending on the bar, you might run into my drinking buddies.

Me, I'm never quite sure where I stand in such distinctions. Even if I'm one of the local assholes, I aspire to something else again.

Big hick town or no, and aside from the pocketed sectarian dementia you can find throughout the U.S., Syracuse seems to have been spared the more lasting effects of the religious fundamentalism you find in some smaller towns. Perhaps this is owing to the influx of European immigrants. Or perhaps those brand-name, Made-in-the-U.S.-of-A. loyalties attached to Motor Town, Battle Creek, Bethlehem are substitutes for spiritual convictions—I can't be certain.

But of this much I am certain: Onondaga County demographics for the fifties and sixties indicate extremely high rates of cancer among men, part and occupational parcel of living and working in the Northeast and a few regions of the Midwest. Even strong men rust, especially in the R & B—the rust belt.

Friday nights after eleven pm, my father drives us over to Syracuse Bowling Center, on the west side. My Uncle Frank is usually there, after leagues. He gives us a few tips. He throws a helluva hook with that shrapneled left arm of his.

First game thirty-five cents, second game a penny. We bowl a dozen each, till one in the morning. Frank and my father drink beer. It's so thick with smoke the pins are hazy.

We join a junior bowling league at Strike & Spare, a few blocks away from Raphael Ave. Mike, Paul, Paul's cousin Vic, and me. Seventy-two lanes, every Saturday morning. I'm the anchor. A good bowler, I tend to pull the ball, but manage some solid Brooklyn strikes.

My mother talks me into skipping bowling one week to take the Pi Mu Epsilon math scholarship exam up at Syracuse University. Determined not to miss even a single week of bowling—fun is tough to come by—I nonetheless begrudgingly agree.

The university: I don't recall ever talking about it before high school, except perhaps in reference to its football team, which appeared regularly on my father's sports parlays. Did my father avoid that part of town on our Sunday drives? I recall the boxy architecture of any number of factories, sure, the State Tower Building, the dome of Assumption Church, but I can't say that, as a kid, I'd ever set eyes up close on the spire of Crouse College. No doubt the "winding curve"—what we used to call that portion of Seneca Turnpike as it curved downhill from west to east into the city, with a steep drop on one side—was a lot more fun for youngsters.

Did my mother ever talk about it?

I've gotten other encouragement to attend college, some in the form of a friend from Liverpool High, Keith Prince. Bright, confident, and popular, Keith is coxswain on the varsity crew team. Headed for Yale, headed for law school, his future seems secure. When he asks me where I'm thinking of going to college—we're both in the National Honor Society—I shrug.

Don't know if I am.

What? What do you mean? You should go—you should be applying. What else are you going to do?

I don't know. Work with my father, maybe?—doing furniture? Work in a factory? I don't know.

Joe—you belong in college. Think about it.

My friend Jerry, on the wrestling team, is also uncertain about college. He lets me sit on his Honda 350 one day—the first time I've as much as sat on a bike. I lean it a bit too far to the left to test its weight, the front tire slipping on the gravel. Then the bike slips out from under me, toppling over as I hop off, still holding the handlebars. The impact bends the left rear directional out of whack. Jerry helps me pick the bike up, inspecting the damage. He grabs a hold of the directional and carefully bends it back into place.

Jesus I'm sorry Jerry, I feel like a jerk. It was heavier than I thought.

No real harm done, Joe. It's all about balance.

Yeah.

So what're you thinkin' of doin' after next year?

Don't know. You?

Not sure yet.

My schoolmate Dale, an easy-go-lucky guy whom I've known since forever—he isn't sure either.

Don't know what happened to Dale. I think Jerry ended up in a two-year college.

Julie is behind me one hundred percent.

Go for it, Joe. You're smart, I know you can do it.

It'll be a decade before she finishes her own baccalaureate. But she's always there for me.

My father tries to be encouraging, doesn't always know how to.

You boys need to get that sheepskin—that'll make the difference. They can't take that away from you once you've got it.

Yeah.

Your mother helped me through my high school degree, after the war.

Yeah.

Smart woman, your mother.

Yeah.

* * *

My mother: if it weren't for my mother.

This isn't the sort of life you want for yourself, Joey.

Uh-huh, I know.

You don't want to end up like your father—working in a factory all his life, and *now* what?

Right, well—it wasn't his fault.

No, of course not—his folks could barely speak their *own* language.

Yeah, I know.

Besides—a college education is—is a good thing, it's a better thing—a finer pursuit. It will help to introduce you to the entire world, not just Syracuse. You need this, this is good for you—especially for a person like you, Joey.

Yeah.

My mother was herself a dedicated student, won regional awards in French grammar as a secondary school pupil prior to the war. Her parents were literate people, of a European class and heritage that imposed such literacy on its children. They would never have expected either of *their* daughters to work in a factory.

I'm wondering what it meant for my mother's life that the war interrupted her college years.

The Saturday morning of the test, my mother picks me up early at 501 Raphael Ave., having left her apartment in Schenectady, a hundred or so miles east, well before dawn. She's more excited about this than I am, and she's insisted on driving into town and driving me there, to make sure I get there. We head up to the campus on the hill, where we have a tough time finding a parking spot. When we finally park, we walk around a little prior to the test, both of us impressed with the quad, sensing among the campus buildings the dedicated pursuit of something other than sheer money. *Money talks, and bullshit walks.* But not here. Or so we believe. My mother waits outside as I walk into the testing area.

I spot a familiar face—Pam Daley, a girl I know at high school. Pam and I are friends, have always done well in our math classes together. Around us in the testing area all we hear is talk about SAT scores and the like. We're both a bit nervous, given the company.

After the test, I introduce my mother to Pam.

Hello Mrs. Amato.

When you meet her, my mother is always cheerful, holds your hand in hers for a moment longer than a handshake.

Hi Pam! Well, so how did you both do?
We look at each other.
Well—
Oh I'm sure you both did fine!
Well—
You did, I'm sure of it!
We say our goodbyes. In the car driving back, my mother is still optimistic. Such a nice campus, Joey! You wait and see—things will work out!

Things do work out—sort of. More like dumb luck.
Turns out that Pam and I tie for third. First and second place winners decide not to attend SU and Pam opts for Le Moyne College. So I end up with first place money.

Meanwhile, back at MoVito's, Mike and Greg discover foosball—table soccer. Mike takes to the game right away—he seems to have a special knack for it, probably owing to the fact that he's been a magic buff since before I can remember, has trained his hands and fingers to be quicker than the eye.
We become pals with the foosball king, Art, a guy we knew much younger, who comes into his own thanks to that game. Art's blinding push shot and ball control rival our meager efforts. And we get to know The Greek. He plays the game the old way, spinning the rods madly.
When we start drinking, the bars—the Poorhouse North, the Gin Mill, Crossroads—double as game rooms. We're not into dancing, most of us, not until much later. We listen to the music, drink, concentrate, and sweat. Somehow we can play well, half shitfaced. And we don't worry about getting laid. Maybe this is what the shrinks call *displacement*.
Fuck the shrinks. Whatever it is, we're hard at it.

Comes with the territory. Overcast, always precipitating, or about to. A plate full of surf & turf every now and then to coax you into thinking you're doing OK, even as you're having the living shit kicked out of you.
We read. A region of working men and women, direct and indirect hand-eye coordination. A place where hands, back, and feet receive much attention, worry, diagnosis. My father has a habit of saying *If your feet feel good, you feel good.* I try unsuccessfully to get him to invert this—*If you feel good, your feet feel good.* Either way, feeling good is linked to foot comfort in this city where the Brannock Device was patented, that polished steel sizing contrivance you find in all good shoe stores.

Land of the Barge Canal, land of manufacturing. Formerly land of Hiawatha, of the Five, then Six Nations.

We read. Grow up playing hit-the-bat, sons and daughters of an era ending. A whole new ball game in the works. Global competition raises productivity demands, "motivating" union workers to rearticulate their commitment to the corporation. Some will come to understand that the salary ranks work for the company too—and salary ranks will come to understand that their collective ass is on the line as well. Everybody is in this thing together, and it ain't a pretty thing. Just drive around Electronics Parkway a quarter-century later, and see firsthand what a quarter-century of layoffs means for local readers.

It'll all but break the unions—*everyone* scared shitless—and open the door for uncontested mergers. Break-time jumping jacks will make a momentary comeback on the nightly news, even as entrepreneurial fervor—read, *more fervor*—will spread across this lacerated land, unemployment rates to rise until five-buck-an-hour service work will look goddamn good. Will even look like, hell, a career.

But us, then—all we hear about is a decent day's wage—if you're lucky, you earn it, and this is what unions are for. That's what our fathers tell us anyway. Our mothers generally nod in agreement—they want better for us.

Use your head, go get a job.

Use your head, go get an education.

Most of us try to do both. Most of us have to.

My father grew up downtown, has little time for hicks *or* religious fanatics. Still, he has one good hick friend at General Electric, Bruno. Nice, gentle guy, built like a fucking ox. Works a mud farm up north a ways, out in the sticks. Great mechanic.

Truth is, the guys I hang out with, like me, have one foot in the city, one foot in the suburbs, and their wheels in the country. Within a fifty-mile radius of Syracuse is some beautiful country—the spring-fed, crystal-clear Green Lakes, Jamesville Reservoir, Oneida Lake, Lake Ontario, the Finger Lakes, a number of fished-out rivers and streams—Seneca, Oneida, Oswego, Salmon, Butternut, Lime, Nine Mile—that manage to retain some charm. And of course the huge Adirondack Forest Preserve, within a couple of hours drive, overdeveloped in many places, still wild in many more. If you're not careful, you can walk in and never walk out.

And then there are the waterfalls—Chittenango Falls, Pratt Falls, Salmon River Falls, Buttermilk Falls. We'll drive out for the day to stand at the bottom of the cascades, throw a Frisbee around.

One small waterfall few know about, off of Route 92, just outside of Manlius. You can walk under, nearly get behind it. I used to catch a glimpse of it as a kid, riding the small choo-choo train that circled around the old amusement park, Suburban Park. We stopped going there when black folks started frequenting the place.

It's all niggers now, boys.

Now, all that remains of Suburban Park are a few concrete foundations.

Not much to do winters but watch the world on TV, listen to it on the radio. Or go out and drink, party. Shoot the shit over booze, pot, whatever. That's six months of the year. And not much to do the other six months either.

Problem is that when we drink, most of us, we drink more than we can hold. The results are predictable.

At the annual collegiate regatta on Onondaga Lake first Saturday in June, twenty-five thousand of us show up ready to let loose, all set to vent that intense energy pent-up after a long, hard winter indoors. At one regatta, a gang of drunken he-men decide to spend the day ripping off the tube-tops of women passersby. Naturally, a huge fight ensues. The riot squad arrives, shields and dogs employed to break up the fun.

The Liverpool Field Days are eventually shut down because of too many fights.

And the Moyers Corners Field Days, the biggest of its kind in the area, always features a Liverpool–Baldwinsville brawl, in or around the beer tent. Sometimes it's a catfight, as we call it then.

Fuckin' Bville, y'know. But they say the same about us Liverpudlians.

So maybe our behavior, all told, is genetic by now, a bad-weather wisdom. May as well be. Some of us will never recover from the place, from its stories.

Before the bar scene we hang out in the major arcades, playing foosball.. Games Galore, Tilt & Tally—the crowd, mid to late teens, follows from hot spot to hot spot. It's a quarter a game, and the quarters flow for five or so years. This same crowd gradually ends up in the bars. Mostly guys. Us. And a few women willing to hone their talents to pass an odd rite of passage: *Put up or shut up*. All of us growing older by the minute, by the second, oblivious. But having sleight-of-hand fun, manually.

Gimme a Molson.

Keep the table.

(The ultimate generosity, as Mike puts it, offered after some six bucks' worth of wins, to novices who haven't yet grasped the flows.)

We take a game made for two and play singles. We take the potential chaos of play and give it a clinically sublime, and sublimely absurd, order. Talk eight ball or surf the asphalt all you want: for the time being, we're playing the only game in town.

Even when we're old enough to drink, we still hang out at Tilt & Tally. The foosball area is all business, and features the table of choice, Hurricane. This is a game for control freaks, and control freaks want well-lit tables, and a layout that allows for plenty of standing room, and room for that quick shift-kick of your right leg-foot that usually accompanies your shot at goalie. At Tilt & Tally, the Hurricanes have their own room, isolated from the sounds and lights of the video games. A few speakers pipe in top 40s rock, saturate the space with the only good times some of us will ever know.
Dreamweaver or high-heeled boy, rock on or truckin'. Either way, FUCK YOU.
Please take your fucking hand out of the fucking goal, asshole. THANK YOU.
Foosball will be arcade's Arcadia, chief emblem of a lowbrow culture's last, non-electronic stand. My brother will self-publish a book, *How to Play and Win at Foosball*. I'll write the foreword, and dangle a modifier—*With this in mind, this manual* . . . (This is a couple-three years prior to my brother and his buddy Matt's *How to Pick Up Chicks with Bar Top Tricks*. You gotta hand it to those two for trying, though.)

We read, reading, then write, writing: the foosball furor will die down a year or two after W. T. Grant's files for bankruptcy. What kills it is intense specialization coupled with a certain lack of standards. Tournament Soccer moves in with their brand of play, and with their brand of table—solid rods, chubby rubber handles, and textured balls—which renders many of our Hurricane-based, flick-of-the-wrist moves extraordinarily difficult, or inefficient, or . . . obsolete. After TS establishes a professional tournament following, the game pretty much dies in the bars.
Today, owing in part to sporadic tournament play, the game has made something of a comeback, a stubbornly solid holdover from a bygone mechanical era. You'll see the occasional home item (like the prop in *Friends*), and some sports bars even host half a dozen (usually Tornado) tables. But you'll rarely find a shoot-from-the-hips crowd humble enough to put up a quarter. That hungry, unregulated gaze?—that gaze is reserved for us long-in-the-tooth fanatics, diminished now to coterie proportions. I helped Mike heft an expensive table into his basement recently, and Art,

now married and living not far from Disney World, is still doing the tournament scene.

But it's a different ball game.

Kids today, say the kids of yesteryear, come by too much too easy. Then again, you do what you gotta do, you can do. And that's the way it goes, right? Some adapt to survive. Some shit or go blind.

Shit or go blind.

Specialists aside, this frenzied, hunched display of Put Up or Shut Up is apt to prove too Who, too sonic boom, too boozy and overheated to foster the poker-face attendant to good corporate breeding. This helps to explain why keyboard cowboys often lavish a table on their cubicle spaces: it's a patently retro exercise in making old decimals speak defiantly, if fetishistically, to the vitamin-enriched 21st-century office classes. But that's another story—of a latter, more volatile, even less stable working-class day.

And anyway, slamming those balls around, I could give a shit about good corporate breeding. Or explanations. Or another story, kid.

One winter night, after the Camaro accident. We're driving Julie's beat '68 Chevelle, driver's-side door smashed in. She wanted to give it to us, my father insists on giving her a hundred bucks for it. You have to get in and out the passenger's side.

The windshield wiper motor burns out just as the snow starts to fly—when we don't have the bucks to replace it.

To put it another way: it's not a priority, like food. Plus, it's a bit of a bitch to get at to remove & replace.

So, we knot a piece of packaging twine from one wiper to the other, in and around through the small corner windows, forming a loop. We operate the windshield wipers manually. While driving.

We get used to this arrangement, so we run it this way the entire winter. Bastard when the string breaks, in traffic, and it's coming down. Easy to get nervous, lose control.

That crate develops quite a following.

Hey you guys, I can't believe you guys!

It's Larry, he's laughing. You can see the glue stains from where he glues his teeth back into his mouth. Bad teeth—really bad. And I should talk. Last time I'd visited the dentist—first time in years—I'd had no less than eight cavities. I'm concerned about this—my father's side of the family are all toothless by the time they turn fifty, no shit.

Larry and me, gotta be genetic.

Mike and I get out and walk with him into Tilt & Tally. It's snowing, about an inch an hour.

* * *

So I'm standing alongside Mike in Tilt & Tally three-four hours a night, sweating, losing quarters. Usually I play back, and he's up front, with his unfuckingbelievable pull shot. Yeah, everyone's got a pull shot on a Hurricane, I know. Just like everyone's got a red sauce, till you taste a good one. Till Mike slams that ball past your taste buds, armless men torqued to life by prosthetic wrists, the melding of sinewed motive and plastic embodiment hitting home with somewhat less delicacy than a metaphor, mixed or otherwise. Score one for the ludic.

Sometimes I just watch him with another partner. Got such a subtle *and* unsubtle grip, Mike actually breaks plastic foosball men in half when the ball wedges (he collects their remains in a shoe box). If you want to block his pull shot on defense, you can't blink. And you have to know beforehand where it's going.

Once in a while we switch positions—my push-kick shot, when I'm on, ain't half bad. When we play each other, and even though he beats me more times than not, Mike insists that my play profits from an inordinate number of "shit shots." (He includes an entry for the shit shot in his book. "The possibilities are limitless!" he jokes.)

Sometimes, instead, we head over to Dolores Terrace, usually to hang out with Frank, Greg, and Rick. The five of us sit in the car, parked at the end of the street, in front of the Beckers'. The radio is always on as we bullshit. I start it up every now and then, turn the heat on to keep it warm. Usually Mike or Greg lights up, and everyone tokes for hours on end.

Everyone but me. My lungs can't handle the smoke. And on those rare occasions when I *do* smoke, it speeds me up. Once, high on hash oil, I grab a straw broom out of Julie's garage and sweep the entire end of the street.

Hey, crack a window will ya?

Rick cranks the back window an inch, the smoke pouring out into the cold.

Sometimes I head inside Julie's house, and those guys take the car, stoned.

On that particular night when we meet Larry outside of Tilt & Tally, Mike and I leave around eleven. It's dark, snowy. We get in around half-past, trudge up the steps.

My father is sitting on the couch in the dark, smoking, drinking a bottle of Carling, muttering to himself.

The TV is off. Bad sign—really bad.

No good two-timing whore.

Mike and I have learned not to press him when he's like this—no telling what might happen.

Four years prior, we walked in one night to a similar scene.

We say hi, he barely responds. We tiptoe around the apartment, getting ready for bed. We're not ignoring him, exactly, but we're not engaging him either. We know he's eventually going to turn his attention on us. It happens when we're just about to go to bed.

Boys—come over here!

We trot over.

Yeah Dad.

Listen, I wanna tell ya something—if that NO good TRAMP Peachy should ever call here again, I want you to tell her that I'm not home! and that I don't want to see her and her little SHITHEAD son! and I don't want her to call here again!—OK?

OK.

OK, yeah.

—you shoulda seen what I said to her, I finally caught her lying to me, that no good two-timing WHORE! So I ended it, I ended it! Like your mother, who was unfaithful, turned into a no-good tramp with that asshole Henry Poster who wouldn't do the right thing, wouldn't marry her, and I showed her too and I punched him right in the fucking MOUTH—

Mike and I are thinking, We've heard this before. It's like this every other week. I'm usually the one to try to talk us out of this situation.

Dad, we're tired—can we go to bed?

—I told her that I didn't wanna see her . . . TRAMP face around here anymore! After everything I've done for her, and her . . . CREEP son! Shoulda drowned that little BASTARD at birth! Imagine giving birth to that . . . THING—

Dad—

He's talking to himself, as though we're not even there.

—after all I've done for her, helped her out with her car, with her job, with that fucking BRAT of hers! And she cheats on me with that other WHORE, Tina! Imagine that? BITCH! Once a douchebag, always a douchebag!—

Yeah, OK. Can we get some sleep?

—you listening to me?—

Yeah, we *said* yeah. We've been standing here for ten minutes. Why don't you just go to sleep?

—you little sonofaBITCH JOE!—

My father hurls his beer bottle at the living room wall, smashing it to pieces.

Mike and I walk instinctively into the bedroom. He follows us in, furious. He picks up the baby-blue dresser in front of my brother's bed, raises it up in the air and brings it down on one end, the entire dresser crushing under the impact, legs breaking off and rolling under our beds.

He yells at us for another few minutes, at the top of his lungs. He's seething now. We don't say a word.

When he leaves, Mike looks at me.

Real good, jerk.

The glass fragments are still visible in the living room wall when Tilt & Tally is making a fortune on our quarters. Around that time, my father happens upon a guy who's selling an old pinball machine. Not some cheap imitation—a full-size, working, arcade-quality pinball machine, complete with cowboy motif artwork.

So my father picks it up for fifty bucks, and Mike and I help him carry it up the stairs at 501 Raphael Ave. We set it up in his bedroom, coin-slot end sticking out in front of our two-foot diameter dartboard. We adjust it so there's no *way* it'll tilt.

The dartboard has been hanging on the wall for years. The drywall surrounding the board has a ring of pinholes around it, extending outward in decreasing density to form a concentric circle some three feet from the edge of the board.

And out on the porch, my father has built a wooden frame from which I hang my Everlast speed bag. Good workout, but it echoes throughout the entire neighborhood.

Jiggling the pinball machine on the wooden floor creates quite a racket itself, especially with the bells and whistles of the machine itself. But as a rule, Freddie doesn't complain—he's learned not to.

Hey Joe—sounds like you've got a pinball machine up there?

Yeah, picked it up for the boys.

When we get tired of pinball, we throw darts. Hard. Or if it's daytime, I slap my speed bag back and forth. Hard.

Anyway, on that particular night when we meet Larry outside of Tilt & Tally, after Mike and I arrive home to my father sitting on the couch with the TV off, we say not a word.

And for a change, nothing happens.

When we get up the next day, we're all a bit silent, subdued—it's like we all have a hangover.

After a few hours, my father asks us if we'd like to play a football parlay. This is his way of saying he's sorry. We say OK.

My mother complains now and again about her first years in the U.S. For one, about the treatment she received at the hands of my gramma. My father backs her up on this.

But there's one thing about which my folks never agree: my mother maintains that my father—that after the war, he'd spend all of their spare money on bets of one sort or another—horses, parlays, what have you.

Believe you me, your father was gambling away our money playing pinball with your uncle Frank. I caught them at it. And here I was slaving away at Kemper Insurance all day.

Between the war, and her work experience in the U.S., my mother has become a frugal woman. Not cheap—frugal, and generous with Mike and me. A single woman earning a modest wage as a receptionist, my mother knows as well as anyone that money doesn't grow on trees.

My father—my father doesn't seem to grasp the gravity of my mother's allegations. He talks about those years with nostalgia. The time he got into a fistfight with two guys, one of whom nearly knocked Suzie off the sidewalk as they strolled by. How she leapt on the one guy's back, dug her nails into his face. And during the war, in Le Havre shortly after they were married, how he and Suzie were staying in a dangerous part of town, how he pulled out his .45 when he heard a group of GIs in the hotel corridor, up to no good.

But when it comes to money, your mother—your mother gets too nervous sometimes. I played pinball with Frank, yeah—why not? It was only a few bucks.

Now my uncle Frank himself—my uncle Frank has always run hot and cold. It's his relationship with my aunt Mary, his wife. Or so I've been led to believe. Something about my aunt having had an affair with a black man while Frank was in Europe.

Not that Frank didn't screw around himself. Who knows? Mary—my mother always says Mary was a "sweet girl." I think my mother could relate to Mary's outcast status. To me as a kid, my aunt Mary seems friendly, and a little loony. These days I know better—looking back, I can see the damage.

Frank himself has had a mistress since the war. And far as I know, he's been playing the horses at Vernon Downs forever.

But my father's relationship with Frank has always been a vexed one. That shrapnel in Frank's left elbow—my father always insists that Purple Heart was Frank's fault. He didn't phone my father, as they'd evidently planned, when he landed in France. My father was a corporal in the Signal Corps, and had somehow arranged, in between official duties and black marketeering, to get his older brother out of combat duty. Didn't work out.

Occasionally my father, Mike, and me are invited over to my aunt Mary and uncle Frank's flat. We show up, knock, but they won't answer the door. We can hear their little Chihuahua, Mickey, barking. Mickey trembles all over when you enter their apartment, his feet ticking and sliding across the wood floors. He lives to be nearly twenty human years.

Thirty years after the war, my father is still playing parlays. Once in a while he lets Mike and me play.

As for my mother, she remains adamant.
Mark my words—that man will never save any money. I had to scrimp and save to buy anything. How do you think I got our luggage? Why, with S&H Green Stamps, that's how! Who do you think planned our trip to the New York World's Fair? Me—your father never wanted to go anyplace. And I had to save every dollar or your father would have tossed it away on—
On what, Mom?
On booze!
Oh c'mon Mom—Dad didn't drink that much then. A beer every now and then maybe, but that's it.
You don't know.
You're not being fair.
But you don't know—you weren't even born yet.

Decades after I first hear my mother make these accusations, my father will tend to spend a good portion of his day sitting in a local tavern, surrounded by pasty-faced drinkers like himself. He'll have lost all his teeth by then, and he won't have the money to replace his old partials. Somebody will say the wrong thing to him, and he'll take his partials out of his mouth, slamming them to pieces on the bar top.
You wanna play around with me?—c'mon, I'll cripple ya you sonofaBITCH—
The two men will hit the floor hard, my father swinging and making contact all the way down.
I do mark my mother's words. I figure, all told, that it must have taken my father some time to develop a sense of responsibility. I figure that—in his twenties, with the war, and his pretty and talented wife; in his thirties, with me, then Mike, and the house at 112 South Dolores Terrace; and through his forties, with a full-time job and real benefits doing what was for him mean-

ingful work—I figure that these all helped to make a responsible man out of him, a homebody. I figure that there's no shortcut to the whole man—I figure that this sort of transformation takes time.

And I figure that what takes time, time can take away.

Not always, of course. As my father would say, *not that sheepskin*.

Still, I don't always figure right, either. Especially *not* when it comes to my own education. You see, when I'd applied to SU, I was accepted into the School of *Business*.

This seems perfectly appropriate to me at the time. In fact, I'd talked it over with my mother.

So you can get a business degree, and major in math?

Yeah I guess so.

Well that might be OK—. I know some very smart MBAs here at CR & D.

MBA?

Master's degrees in business. Some are quite smart.

Yeah?

But what about the sciences? I know a lot of scientists here, and you're good at math. And when you were little, you always wanted to be an astronomer—

Astronaut first.

Yes, but remember all of those astronomy books I bought you, Joey?

Yeah.

They have a man here working on the beginnings of the universe.

The beginnings?

Yes. World-famous scientist.

I dunno. I'm not sure what I'm gonna do for a job, and I figure if I can get the math background, and the degree in business—

I see what you mean. Well that sounds OK then, Joey.

I'm pleased that my mother agrees.

I become fond of the little admission card I'm sent, with the School of Business logo imprinted on heavy cream-colored stock. When my mother sees it, she's excited. My father holds the card between his fat fingers, looking at it and shaking his head.

That's really something.

My father drops me off up on the hill for freshman orientation. I walk around campus, lost, eventually find Slocum Hall.

Following handwritten signs, I end up in a largish room crowded with perhaps four dozen people my age, seated helter-skelter. The room has that sweet gymnasium-sweat smell that seems to seep into the wood floors of old structures.

I take a seat toward the back, in one of the combination desk-chairs. A woman who looks to me to be in her thirties stands at the front of the room, busy talking with a somewhat younger woman and man, dressed in jeans like me. The older woman is dressed in a gray skirt, white blouse, and gray jacket, her hair is up in a bun, her manner is professional, warm, friendly. She brings the room to order.

OK, let's get started. My name is Professor Stanley. I would like to welcome you to the orientation session for first-year students in the School of Business. Everybody here should therefore be a Business major. Is there anybody here who is *not* a Business major?

Confused, I raise my hand—slowly. I don't want to, because I'm finally feeling a bit comfortable. Mine is the only hand raised, and everyone turns to look at me.

OK. Will you please come up here for a moment. The rest of you can get busy filling out the personal data sheets.

I walk to the front of the room feeling everyone's eyes upon me. The younger woman and man begin handing out forms.

Hi Professor Stanley, I'm sorry but I'm—

And what's your major?

Well—I'm a math major.

A *math* major?

Yeah.

So—you should be in Arts and Sciences, then?

Well, I was going to major in math in the School of Business.

If you're in the School of Business you must major in *business*.

Well, I'm on a math scholarship, and I was accepted into the School of Business.

She pauses, sizes me up quickly. A look of genuine concern cross her face, a look I've seen before, sometimes when I've tagged along with my father to Social Services.

First things first: what's your name?

I tell her. She turns to pick up a list on the desk next to her.

Well, here you are, right where you should be. Now, what's the name of your scholarship?

I tell her, she writes it down. She walks over to a nearby desk, makes a phone call. Waiting on hold for a minute or so, she rolls her eyes at me to say she's getting the run-around. This relaxes me some.

Uh-huh. OK, I see—thank you.

She hangs up and turns to me.

I'm so sorry—but you can't use that scholarship in this school.

I'm caught up short. I pause for a moment, thinking, not sure *what* to think.

But I was admitted into the School of Business with this scholarship—

I know, I'm sorry—that was *our* mistake. But you'll need to find another major.

How do I do that?

I'll give you a copy of the undergraduate bulletin. You can look through it—take your time—and when you find another major, you'll need to change your major. But you'll have to find a degree program that your scholarship will support.

How do I change my major?

You'll have to go over to the Administration Building, and fill out some forms.

OK—OK, thank you.

You're welcome. I'm sorry about this—please let me know if you find something. I'll be happy to phone the scholarship office again.

Thanks.

I take my seat again and flip through the catalog. I look up Arts and Sciences, and I spot the math major. I keep flipping through, and notice the College of Engineering listings. I see the various engineering majors—mechanical, electrical, civil, chemical—my eyes jump back to mechanical. The word is familiar.

I stand up and walk back to Professor Stanley.

Excuse me, Professor? I think I may have found something—

Let's see.

She phones again.

Uh-huh. OK, thank you.

She turns to me, smiling.

Mechanical engineering will be just fine.

My mother is excited.

I think you're better off, Joey. I had a hunch you might get into engineering!

My father is happy as long as I'm happy.

Engineering is a good field, Joe. Pays well. Just don't let yourself become one of those smartass time-study dummies!

※ ※ ※

Julie is pleased.
Engineering—that's great, Joe! You deserve it.

Me, I'm not sure *what* to think. I never planned any of this. I like math, but I'm not sure what you *do* with a math degree. And I don't know any engineers, so I know squat about what an engineering job is like.

Two months later, browsing through the catalog again, I spot a dual-degree, five-year program in mechanical engineering *and* math. To a young guy like me, it sounds—reads—like just the thing *for* a young guy like me.

Read it and weep: by semester's end, bombs dropping on Southeast Asia like never before, here's a young guy like me trying to cover his bases, homegrown stars in his eyes.

5.

The Flying Pork Chops
and Other Adventures in Craft and Cuisine

> If it is true that there is no greater sorrow than to remember a happy time in a state of misery, it is just as true that calling up a moment of anguish in a tranquil mood, seated quietly at one's desk, is a source of profound satisfaction.
> —Primo Levi, *The Periodic Table*

OR AS JOHN GARFIELD puts it as Mickey Borden in *Four Daughters*, "Talking about my tough luck is the only fun I get."

I grew up with compressed air. And you could say my father missed his calling.

On my eighth birthday, my mother takes me to the thrift shop. We're to find a desk as my birthday gift, something I can write on.

I spot an old, junior rolltop. Solid oak. But creaky, dark brown, sticky with age. Ten dollars.

The sales clerk helps my mother get it into the trunk of our car. We drive it home, slowly, its legs sticking out.

When we get it home, my father grabs a hold of the desk, lifting it out and carrying it into the garage through the full-width screen doors he's made for summertime. The desk seems to contort as he grabs hold and lifts it.

He sets the shaky structure down. He circles it, eyeing it carefully now, sizing up what must be done.

The desk is put together with horseshoe nails.

He pulls out the nails—all of them—completely disassembling the desk. He removes the lock hardware. Placing some newspaper on the garage floor, he brushes each piece in turn with a powerful industrial-grade stripper to melt the old varnish, using rags and a number of different scraping tools to leave only the faintly stained oak beneath. Then he sands each piece lightly to remove all trace of the varnish.

With his spray gun, he applies several coats of lacquer sealer, sanding each coat in between. Finally, he finishes each piece, first one side and then the other, with a few coats of clear lacquer. All of this takes time—perhaps a week of nights after work, and a weekend. When he's through, each surface is smooth, glassy, the color now what a finisher like my father calls blond.

He has to construct a new bottom for one inside drawer. And he has to fiddle with the locks some, to set them in place better.

All that remains is to drill holes into each piece, fitting dowels on opposing ends. He reassembles the entire desk with dowels and plain old Elmer's glue. He taps the pieces back together with a rubber mallet, and with a damp rag wipes off the glue that squeezes out as he clamps each piece.

Some touch-up follows, the old desk now newly old.

Three weeks later, it stands, shining, in my bedroom.

It shines to this day.

But you could say my father missed his calling.

He has a certain irreverence for custom, and for his craft. While he's working at Hanson Interiors, it's the Italian—him—against the Jews. The Hanson family is Jewish, makes a shitload of money importing Scandinavian furniture. Thing is, the pieces are shipped stained, with the lightest of oil finishes. Customers are furnished with cans of oil, told to apply a light coat every now and then to keep the wood sealed.

My father thinks this is a horseshit way to treat furniture. It's not oil finishes he's opposed to—a true oil finish takes a great deal of time and effort—but the practice of shipping furniture overseas with little protection. The light oil all but dries out by the time the pieces arrive at Hanson's, the grain of the teak or rosewood sometimes raised as a result—my father calls it *napping*, a phrase typically associated with fabrics. Not an especially pretty sight, but this doesn't lower the price any. After all, you have to have an eye for such things.

So my father takes a chance on one piece. He hits the surface with a light coat of clear lacquer. He uses spray bombs. You wouldn't think it would work. You'd think the oil, immiscible with the lacquer, would produce fisheyes. But no fisheyes. The wood regains its luster. And some protection against stains.

Hanson isn't happy about this.

What do you think you're doing?

I'm doing what I know how to do, and what you *don't* know how to do. What you pay me for, right?

Hanson shrugs and walks away.

In the end, my father's idea catches on. The customers are happier when they're told about the "process." The furniture looks nicer, feels nicer when you run your palm along the surface. Eventually, the factory in Denmark learns about my father's innovation. Pieces start arriving with a light coat of clear lacquer.

One afternoon I come by after school to pick up the car.

Joey, I'd like you to meet Mr. Hanson.

We shake hands. Hanson is a not unhandsome man in his mid-thirties. My father tries to be respectful of the well heeled, deferential around his superiors. Cautiously polite, at least publicly. He understands that powerful people can hurt you. What he says behind his boss's back is his business.

A few moments later, Hanson returns. He looks irritated. My father's workplace takes up the back portion of the stockroom. There's a fan installed for fume exhaust when he's spraying.

Joe!

Yeah?

Listen—didn't I tell you to get that piece unloaded and set up? What do you think I pay you for?

JOEY, wait for me in the car!

My father's face is beginning to turn red. The veins in his neck have begun to show. I leave. Quickly, instinctively—I can almost hear what's coming. I sit in the car, and wait.

A few minutes later, my father comes out of the building's rear door, shaking his head back and forth. He gets in.

That rotten fucking bastard JEW! I told him if he has a problem with me, he doesn't read me out in front of my son. I told him he can stick his job UP HIS ASS!

Take it easy, Dad.

TAKE IT EASY? Why that rotten fucking bastard JEW!

He's mad as hell now. I wait a moment. He cools down.

What're you going to do?

I'm going next door to Soo-Lin's, for a drink. I told him I'll be back when I'm GOOD AND GODDAMN READY! You go pick up Mike. Come back and get me around normal time, six.

When I return, I can tell my father's been drinking. He's not drunk, but he's been drinking.

* * *

One of the guys at Hanson's, Phil, has a metal plate in his head. Don't ask me how it got there. Phil is one of the movers, and my father and he don't get along. To put it more precisely, my father doesn't tolerate *anybody* but his boss telling him what to do. And even then. And Phil—well, without always meaning to, Phil has the unfortunate habit of taking orders and communicating them. *As* orders.

I told Hanson to keep that crazy bastard away from me, right Joe? So I'm putting together this knockdown piece, and Phil walks in. And you know what that NUT says? He says Joe, *Nathan* wants you to work on *that* piece. So I tell him I don't give a shit *what Nathan wants*, mind your own fucking business, you NUT. But he just won't back off, Joe. *But Nathan wants* DON'T FUCK WITH ME *But Nathan wants* he just won't leave me alone. So I pin him up against the wall with my right hand, I've got my left cocked back, and I tell him if he doesn't stay the FUCK AWAY FROM ME I'M GONNA BREAK YOUR FUCKIN' JAW! They have to pull me offa him the fuckin' NUT!

My father is flushed, reliving the scene. It's like he's mad at *me*, telling me this story.

Hanson has a young, ten-year-old daughter, Abbie. My father has always had a way with little kids—they seem to respect him, and he treats them well. Abbie is no different. Whenever her father brings her into work, she visits my father in the stockroom. Sometimes my father shows her how to do a burn-in—touching the flat end of a hot putty knife to a lacquer stick and smoothing the blob of melted lacquer into a dent or nick. Then feathering the surface with the hot blade—my father sometimes licks his thumb and runs it across the burn-in to provide a little extra lubrication for the knife. It sizzles on the wet spot.

She's a smart little girl, Joe. Very well behaved. But her father spoils her. The other day I'm frying up a pork chop sandwich on the small electric griddle they have in the back room. Abbie walks in. *Hi Mr. Amato. What are you cooking?*

He smiles, then continues.

Now you know, Joe—Jews don't eat pork. So I offer her a bite.

You did? Did she eat it?

Sure, she loved it, that squirt!

I'm shaking my head, but I'm smiling.

Good little sandwich. And just as she's swallowing that bite of pork, her father walks in. *Abbie, what are you eating? Pork chop sandwich Dad*, she says.

Mr. Amato gave it to me. It's really good! You shoulda seen Hanson's face Joe, I'm not shittin'!

My father is laughing now. So am I.

Three years after my father starts working for him, Hanson tries to screw my father out of a month's vacation. My father quits, and applies for unemployment, explaining his situation. The state gets involved, and when the smoke clears, they award my father his vacation pay, plus back pay. But it's really rough going there, for five or six months.

My father's irreverence for the conventional, his knack for devising tricks of *his* trade — of his own doing — reaches back into the fifties. He learns on the job, using only those techniques he finds useful.

He's the first finisher working on the line to use airbrushes for touch-up. He becomes an expert in color patchwork, burn-ins, repairs of all kinds. Sometimes he burns-in a nick or dent, matching the color as closely as possible. Then sanding, or sealing and sanding. On smaller damage, he sometimes feathers in the grains with a good camel-hair brush — no camel hair, actually, it's just what they call them — by hand. Then he hits the surface with a coat of lacquer. All of this as fast as you can imagine — he's on an assembly line, and when too many damaged cabinets back up, somebody is there, watching.

It takes know-how to persevere in this line of work. And a thing more. Passion?

But for all of his years at General Electric, he can never make it into the finishing shop, where the initial finish is actually applied to the TV cabinets. The men there have more seniority. Some are old-timers. Most of them still apply stain with a rag. At home, working on the damaged furniture he repairs on the side for Allied Van Lines and others, my father applies stain with a spray gun.

By the time my father takes his last full-time job up in North Syracuse, at Ned Popoff's Organ and Piano, the hazards of his craft, combined with his smoking and drinking, have begun to take their toll. He's hired by Popoff's to replace their chief finisher, Ralph Stark.

Ralph Stark: one of the old-timers in the General Electric finishing shop. Pushing eighty now. When they meet, Ralph remembers my father's name, but he doesn't remember much else. And my father keeps to himself pretty much, so it's a while before they get to talking.

You see, my father doesn't have time for small talk, doesn't fraternize much until he gets to know you — he wants to make sure you're not an asskisser. On my father's rather extensive list of things in the world that you should hold at arm's length and in utter contempt, asskissers are right at the very top.

Respect for the boss—sure. But respect is a two-way street. An honest day's work—of course. For an honest day's pay. Asskissing—you must be joking. Grateful is, at most, grateful.

To his credit, Ralph turns out not to be an asskisser.

One day my father is hard at work on a damaged piano top. A black, glistening Yamaha, worth thirty grand anyway. The movers have slipped with it, as movers do, and smashed up one corner but good.

My father saws, sands, shapes, drills, dowels, glues on a new corner. He files, fills, masks, seals, sands, sprays—*refinishes*, his arms moving every which way. He's doing things to that piano top you wouldn't dare—pressing down hard with a file, testing and retesting its surface against all the chemicals and materials within reach. He's pushing past mere technique, past the limits of craft, breaking the rules to make that surface yield a beyond that seems both beneath and above.

And all along pushing the limits of his body, lung tissue and membranes inhaling and absorbing these compounds, arthritic left hand—his sanding hand—aching. But his hands are steady while he works—rock steady. He's attending to that wood with the same combination of care and carelessness that any artist brings to her work, working both with and against the grain in so many ways. And all for eight bucks an hour, under the table.

In his concentration, my father doesn't see Ralph, who's standing behind, watching him work. After fifteen minutes, my father hears a murmur. He turns.

It's Ralph. He's welled-up, choking back the tears.

Joe—Joe, I didn't *know*.

But yeah, you could say my father missed his calling.

He's a helluva cook. Ask my mother—*after* the divorce. Ask my girlfriend Julie. Ask Mike. Ask anybody who stops by 501 Raphael Ave. At first, though—

When we move into the flat, my father tries cooking us macaroni and meatballs—like his mother. Those first meatballs are hard as a fucking ROCK. Only thing I can ever remember him cooking when we lived at Dolores Terrace is fried dough. He did a good job with the dough, but meatballs are another matter.

So he decides to ask my gramma, find out how she does it.

The Flying Pork Chops 73

* * *

Now my gramma, Antoinette—actually "Antonia" in Sicily, renamed upon immigration in 1913, as I'll learn a decade after my father dies, searching the digitized Ellis Island archives from our townhouse outside of Boulder, Colorado—Antoinette is a difficult woman, has led a difficult life. She holds her fists against her chest when she's upset, uttering *Jesu Christo* over and over, *Jesu Christo*, along with a lot of words I don't understand, *Jesu Christo, koh-may-see-deech?* Goes to church mostly on holidays. You can call this religion, if you like—she's a believer, that much is certain. Seems a superstition to me, Old World or new. Not faith, but blind faith, the faith of the needy—who, like all of us, need faith to get by, but who, unlike some of us, are provided scant reason to be faithful.

For no apparent reason, my gramma kicks her third oldest son Joe, my father, out of her house shortly after he returns from Europe with his French war bride Suzette, my mother. My grampa finds them another place to stay on the west side, on West Belden Ave., with his sister Rosa—Rose—and her husband Frank Squadrito. This is where I spend the first year or so of my life.

My gramma gave birth to a child who died as an infant—the first Domenico. He's buried in St. Agnes Cemetery, up on Onondaga Hill, at the southwest edge of and above the city.

Rumor has it that my gramma, at some point in the distant past, had a miscarriage. Drank some animal blood or some such once, whether to precipitate the miscarriage or as a healing remedy, I'll never know. Rumor also has it that, also in the distant past, my grampa punched her in the mouth once, knocking her teeth out. My gramma, with all of her complaints about her health, will live to be one hundred and one. She'll outlive her son Joe, but not her other three sons. I'll always remember her bending a garlic clove this way and that with her fingers, to loosen the peel. She had all the patience in the world for this.

My father has a tough time finding out from my gramma how to make meatballs—the recipe is not so much in her head as in her body. He spends one afternoon exchanging Italian and English snippets with her, jotting down the results in his rough uppercase.

Loh skree-vay, Ma please, ah-speht—ree-pay-tah, layn-tah-mayn-tay?
Si Joe, oon poh sah-lay.
Kwahn-toh?
Oon poh sah-lay, uhm—sahlt?
He also gets a general idea of how to make a sauce.
This is his first cooking lesson. He's going on fifty.

But he seems to have a knack for it, for turning a few simple ingredients into a tasty rustic meal. Before you know it, it seems he's cooking all the time, to match his eating habits. He drinks many of his meals, so he's hungry at odd hours. His hands always tremble when he serves you.

Saturday morning: Mike and I awake to the sound of food frying, in the background another hissing sound punctuated by near-perfect intervals.
SSSSshhhSSSSshhhSSSSshhhSSSSshhhSSSSshhh.

It's my father, finishing a piece of furniture out on the porch with spray bombs. The fumes drift into the house, mixing with the garlic-laden meats on the stove. Lacquer fumes and fried garlic, fried garlic and lacquer fumes. The cigarette smoke we don't notice. My mouth waters. We get up.

You boys wanna have a little breakfast?
He claps his hands and rubs them together.
Sure.

He grabs a rag and pours some lacquer thinner on it, cleaning off his hands. The pungent solvent odor we're all used to fills the kitchen. Then he washes his hands with soap and water. In twenty minutes he's fried six sunny-side-up eggs for me, three scrambled eggs for Mike, three links of Italian sausage, and a large helping of home fries and onions. And toasted a half a loaf of Italian bread, under the broiler. Instant coffee (Nescafé) and orange juice, and we're all set.

But we don't always have money for food.

When we first move into 501 Raphael Ave., my father tries to keep rent, gas and electric, and loans paid up-to-date. He finds he has to keep borrowing—from loan companies like Beneficial Finance, from friends. This is long before most men, women, and children in the U.S. are offered a credit card by virtue of their citizenship. Trying to keep up with the bills means that my father sometimes runs low on food money. We're off and on public assistance, receiving food stamps. Along with the standard assortment of social service food: Spam; round containers of powdered eggs; raisins.

Spam is Spam; powdered eggs are powdered eggs; and food stamps— whether you appreciate their value often depends on the cashier, and on the length of the checkout line.

One evening I decide to crack open a box of government-sponsored raisins. Not much in the house to nibble on.

I tear open up the box, and take it into the living room where, in the dark, I lie on the couch watching TV. After five or ten minutes of nibbling, something tastes funny.

I get up and switch on the lamp. When I look down into the box my stomach does a flip-flop.

LICE!

I run out into the dining room, where my brother is working on a lab report.

LICE!

My brother jumps up, looking into the box.

They're not lice, you idiot—they're *maggots*!

The box is FULL of maggots. I drop the box on the kitchen table, run to the bathroom, flush my mouth with hot tap water. Brush. Gargle. Drink three cups of hot tea. Mike is laughing his ass off. Takes me a while to see the humor in this.

My father laughs when Mike and I tell him this story.

But it doesn't take him long to adjust his priorities. It doesn't take anybody long.

From now on, whatever else, you boys won't go hungry.

My father's conviction brings with it a certain stipulation: You don't bitch about the food.

Like my father, I've always been one to bitch.

So one day Brian drops by. My father asks Brian if he'd like to hang around for supper. Brian says sure. We piss around outside for a couple of hours, walk back upstairs. My father is standing at the stove with his back to us, spatula in hand.

Sit down boys.

My father has set the table, a large stack of home fries and a bowl of corn already in place. And a salad—oil and vinegar, salt and pepper dressing—with uniformly chopped chunks of tomato, celery, onion. The three of us sit down at the table. The food on the stove has a familiar smell.

What are we eating, Dad?

Food.

What food?

Pork chops.

As he says this, he walks into the dining room with a fryer full of pork chops, poking at the chops with a fork and beginning to drop two on each of our plates. I look down at my plate. I realize I'm about to commit a serious error in judgment, but I can't help myself.

Pork chops? Again?

This is all it takes.

Yeah, PORK CHOPS! Whatsamatter—you don't wanna EAT?!

I didn't say that Dad—
You DON'T LIKE PORK CHOPS?! JESUSCHRISTALLFUCKIN-MIGHTY!

He's now generating a southern Mediterranean heat zone that's hard to appreciate until you're in it. Not a large man, he seems to grow in size the madder he gets. And with each word he gets madder.

You DON'T LIKE TO EAT PORK CHOPS?! WHY THEN YOU WON'T EAT YOU LITTLE SHITS!

He turns and hurls the fryer into the kitchen sink, the remaining pork chops sailing across the kitchen. The three of us push our chairs back from the table. He turns and jerks the end of the table head high to bring it crashing down on the floor. The dishes glasses home fries corn salad pork chops slide off and up into the air, spilling and crashing all over the room.

We're ducking, every man for himself. My father storms around the kitchen throwing pots pans forks knives spatulas this way and that. Brian stands up, trembling, barely audible.

See you guys later!

He walks-dashes out the door and down the steps.

Bye Mr. Amato.

Mike and I get up slowly, walk into the bedroom. Mike looks at me.

Jerk.

We can hear my father walk down the steps, get in the car, and leave. He comes back a few hours later. We've cleaned up the mess.

Such events are not without instructional value. Twenty-odd years later I'll nearly split a gut reading a simple line in a poem written the year I was born, because I'll goddamn well know who killed the fucking pork chops.

But these are the early days. As we grow older, we learn to adjust some to one another's habits and quirks. Laughter is one way—a kind of glossing over, whatever the damage done. Works all too well when narrating after the fact—we tell these stories again and again to one another, to others, trying to get them just so, picture perfect. But patchwork conceals scars, the worst of times, those senseless silences that butt up against the angry howl of a vowel sandwiched between a couple of unfortunate consonants.

Mike and I fight.

At first, I'm older, and stronger. I can pin Mike down, and all he can do is spit in my face.

But as we grow older, Mike can hold his own. His legs are tremendously powerful, if my upper body is a bit stronger.

On his first day at Liverpool High, I bring Mike up to the universal gym area. The sweaty room is crowded with bulky jocks. My brother spots the leg press, and instinctively sits in the seat. KERCLUNK. With zero effort, he pumps up the 450 lb. setting, using the top pedals. A few of the jocks notice this.

Raise it to the top.

I pull out the peg and place it at the limit—750 lbs.

KERCLUNK.

A few of the jocks stop working out.

His legs are short!—that's why he can do that!

Mike doesn't say anything. First, he pushes the 750 lbs. up on the bottom pedals. Then, to add insult to injury, he gets up in his seat, positions his ass against the red vinyl back and twists his body sideways slightly. He puts *one* foot on the top pedals and, with some effort, pumps the entire 750 lbs.

KERCLUNK.

The jocks all stop, astonished. I'm smiling. My brother has been a tree-climber all his life, once long-jumped 21'6"—barefooted. In junior high, with the coach watching.

But on the high school track team, Mike can't hit the board square. And the coach won't spend any time helping him practice, has his hopes set on the older athletes. So Mike decides to do somersault long jumps, usually nails around 17'.

What this translates to is that you can't knock Mike down. You have to pick him up.

When we fight, I sometimes end up socking him. Hard. He tries the same, but he's quicker with his feet, and you can find yourself on your ass real quick.

I still have a scar under my lower lip from one particularly animated fight. My brother hits me in the face with this small metal calendar—my mother had brought us each one, from Florida. When I look in the mirror and see the cut, I walk back into the bedroom and go ape-shit on him as he huddles up. I know I bruised him but good.

If my father is around, he usually doesn't interfere until it gets pretty rough.

One night we're going at it in the living room. My father is sitting on the couch. It's around eleven.

Boys, I think that's enough.

I'm holding Mike with one arm around his neck, digging my elbow into his side. He manages to swing his double-jointed shoulder around, while kicking his right leg under my legs as he does so. We both go down, hard, on the wood floor.

BA-BOOM.

Boys! Enough I say!

We stop wrestling. We hear a knock at the kitchen door. My father gets up. It's Freddie.

Joe, sorry for coming up, but could you ask the boys to keep it down a little, please? The kids are trying to sleep, and—

OK Freddie.

Thanks Joe.

My father shuts the door, and turns to us.

Now behave yourselves!

My father teaches my brother and me how to cook. He leaves us notes when he has to work late.

BOYS: TAKE CHICKEN LEGS OUT OF FRIDGE. PLACE IN OIL ON STOVE WITH SALT AND PEPPER. LOW HEAT 25 MINUTES. TURN ONCE. PEEL THREE POTATOES. SLICE THIN. PLACE IN OIL AND FRY. WARM UP PEAS. LEMON JUICE FOR CHICKEN. BACK AROUND 7. LOVE, DAD.

All uppercase, slanted—to the right a bit. They forced him to write right-handed as a kid. Me, mimicking him—I sweep and shovel left-handed.

Now when it comes to more intellectual matters, Mike usually sides with my father. I mean, we're all blue-collar Democrats, strictly speaking. We'd all agree, in principal, that it's imperative we find ways to run the corporations without further abuse of the workforce. And none of us gives too much thought to the possibility that corporations might eliminate all forms of life on the planet. Not at this time, anyway, and not in this place—jobs take priority over the environment.

But even if our tactics are often similar—we argue like cats and dogs to get each and every point across—our politics are often different. My father holds to a more individualistic sense of success and responsibility and justice, and Mike follows suit. Me, I'm more wedded to the critique of the corporate welfare state, a bit left of their left.

I attribute this to my being the firstborn, the one upon whom so much initial attention was lavished. More photos of me as a baby, perhaps more sheer wonder in the eyes of my parents. I signify a new start. Which means that my folks' values are likely to become my values by default—if I'm not my own man, if I don't set my own example.

Mike? Food-wise, his being the younger means that he's somewhat less wedded than I am to old-world cuisine—like Greek olives and figs, or pickled eggplant, or squid, or sausage, or on my mother's side, sauerkraut, or onion soup, or kuchen. He's more likely to order a hamburger, more likely to have a yen for a peanut butter sandwich. But in a lot of ways, circumstances have been tougher on him than they have on me. If I'm given the burden of proving myself a free thinker, he's given the burden of not.

Of course, when Mike reads this, he'll tell me I'm *full of shit*. And so I am.

As kids, we're never grounded. My father just raises that left hand of his and we behave. Once in a while he spanks us on the ass a couple of wallops. It smarts, wounds our pride, but it doesn't actually hurt. After which we're sent to our bedrooms for a half hour or so.

OK, you boys think you can behave yourselves?

Years later, when my mother visits from Schenectady, it's my father and Mike who'll often wrestle with each other, playfully, in his north side apartment, my mother shouting at the two of them to stop.

But it's Mike who'll be the first to move out of 501 Raphael Ave. It's Mike who'll have the hardest time dealing with my father's ways, because it's Mike who takes the brunt of my father's anger. Because it's Mike, the younger son, who's learned to identify with my father. Because it's Mike, the younger son, who's felt the pressure of coming-after, of exemplifying.

And this paternal identification, this father-son expectation and intimacy makes Mike more the object of my father's rage, if not more vulnerable to it. They see each other in each other's eyes, and they don't always like the resistance they spot.

Sometimes my father slaps Mike hard, in the face, but this often fails to produce the desired effect.

Go ahead—hit me again, jerk!

And he does. And Mike yells the same thing back. And rarely does it do any good when I intervene, or try to. My brother is stubbornly resistant, my father's anger is resolute.

Shut up Joe.

My father knocks us around when he's mad, when we've misbehaved—or when he thinks so. He stings, but he never really hurts us. Generally we just duck & cover. On one occasion I instinctively double my fists, and he doubles his, two fighters ready to go at it. Fortunately for me, that's about as far as it gets.

My father scares the living shit out of you, in part because he loses control of all but his capacity for inflicting serious bodily injury. He may not hurt you when you cross that fine line of his, but he will most definitely shake you up a bit. And it's everything he can do to focus his rage. He may destroy anything that's within reach—furniture, walls, whatever. You regret contributing to his anger because you know that, an hour or two later, *he'll* regret it—it's in his eyes, the depression that sets in after. Once in a while he'll order a pizza, to compensate.

But here again, I'm talking about Mike and me. Others are not so lucky.

Every once in a while my father likes to tell this story about his years in the Army, in France. He's heavy into the black market, selling GI goods for a profit. Cigarettes, booze—small stuff.

And one day I spot this bitch—*every*body knew she was a collaborator with the Germans—talking to one of the sergeants. She was gonna blow us in.

How'd you know?

She'd threatened us.

So what'd you do?

He lets out a laugh, takes a drag on his cigarette.

Why I knew just what to do. Right after the sergeant left, I stepped across the street, walked right up to her and punched her right in the goddamn mouth, that no good douchebag. Fractured her jaw but good. And when she hit the pavement, why I told her to keep her fucking tramp mouth SHUT! I took care of that shit.

He's got his fists doubled as he says this, almost smiling.

Nice, Dad, real nice.

What was I supposed to do? Let her blow us in?

My father was raised during the thirties on the north side. Rough part of town, and the kids play rough. According to him, they don't play. He's light complexioned, like his father—who, like everyone else in the neighborhood, calls his son a *pollock*.

He ran at me with this stick, and tried to stab me in the head with it. He woulda killed me.

My father is eight, his attacker nine.

And I kicked the living shit out of him.

My father is sixteen. He's talking about one of the DeJohn (Di Gianni) boys, one of whom, Joey, fought Jake LaMotta at the State Fair Coliseum in 1949, giving LaMotta a real shellacking, my father says, before running out of steam in the eighth.

One night, around nine, my father comes in half stooped-over.

Dad, what's wrong?

Guy hit me.
What?
He can barely speak. Mike and I give him a hand, get him over to the couch. He's holding his ribs.
What the fuck happened?
I'm sitting at a bar.
He stops, wincing.
Then what?
I walk in, sit down. Two big guys on my right. Get up to go to the john.
Yeah?
Left their money on the bar.
He pauses again.
When they come back, I can hear one say something to the other about their money.
Yeah, then what?
Next thing I know I'm on the floor, the bartender is helping me up.
What?
The bartender . . . told me one of the guys thought . . . I took his money. On the bar. So he punched me from the back . . . in the side.
Fucker!
Neither Mike nor I knows what to do. My father is not the type to call the cops, not unless he thinks the cops can actually help. And for my father, help in this situation amounts to revenge, plain and simple.

For three months he lives with that injury, won't see a doctor. The contusion is severe, covers his entire right side. It's possible he's got a broken rib or two.
But he refuses to see a doctor. And it's not just because we don't have medical insurance. He's afraid to—afraid of what else they might find. But I can never be certain whether it's the threat of death he's afraid of—cancer, say—or death's repercussions.
One weekend, four months later, Mike and I get in a bit late. My father is lying on the couch, having a beer, smoking, watching TV. He's in good spirits.
Hi boys.
Hi Dad.
Hey, got a minute? I wanna tell you a little story.
Yeah, sure.
My father sits up. Mike sits on the couch with him, I sit in the chair.
So I found out who hit me.
Who?!

Mike and I both jump up.

Hold your horses. Let me explain. I keep going back to that bar until I see the sonofabitch.

Then what? What the fuck happened?

Hold on, will ya Joe! I'll tell you. And watch your mouth!

C'mon Dad will ya?

OK! Just let me explain. So I see the guy—big young guy, around thirty. It turns out he goes there every Thursday night.

Yeah?

Yeah. So the other night, I pull into the parking lot, and wait. I have a knit cap on, pulled down low, covers most of my face. I take a piece of tape and tape the light switch, in the door jam.

So your dome light won't go on?

Right Mike. And I wait, with the car door slightly open. He comes walking out, and as he starts to walk over to his car, I get out. He doesn't hear me coming.

Yeah?

Yeah. He opens the door to his car, and as he goes to step in, I'm right there, with a tire iron in my hand. And I say you nogoodsonofa BITCH HOW'S IT FEEL! And I lay one across his back BUT GOOD! And he falls into the car and I lay another one right across his side, just like HE HIT ME AND I SAY THERE YOU BASTARD I OUGHTTA KILL YA!

Jesus Christ!

Mike and I are laughing so hard we're almost pissing our pants.

What'd he say?

Not a fucking thing. He just lay there, groaning.

My father picks up the *Post-Standard* lying on the coffee table. It's folded open to page six. He hands it to us, mischievous glint in his eye, proudly pointing to a small column with the caption "Man Assaulted on North Side." I realize then my old man, just what a crazy crafty bastard he can be.

Depending on my mood, I generally take charge of a different cleaning operation every few months. Sometimes it's the kitchen entrance, alongside of which my father has a habit of stacking boxes and boxes of finishing equipment, jars full of this solvent or that stain. Sometimes it's a specific room of the apartment—the living room, and my father's coffee-table nightstand.

My father generally lets me have my way, even as he complains. He appreciates my sense of order in the midst of our daily disorder.

But my father has what I view as an unhealthy tendency to hang onto

The Flying Pork Chops

food far too long. And I tell him so. He wraps everything up in aluminum foil, and you can't tell what it is without unwrapping it. We hadn't figured out then that Magic Markers work wonders.

Mike is in the living room watching *The Newlywed Game*. I'm sitting at the kitchen table, my father has just walked in. He opens the refrigerator.
See you guys know what's good, dontchas.
Huh?
You know what's good, dontchas.
What's that?
Escarole.
If we knew what's good we wouldn't eat that crap.
What's that?
If we knew what's good we wouldn't eat that stuff.
Mike and I laugh.
Ate it, didntcha.
We laugh harder.
Ate it, didntcha.
So we almost didn't.
You ate it, you knew what's good didntcha, huh.
He closes the refrigerator, I get up and open it.
Lookit—I suggest you do something with these eggplants 'cause if you don't I'm gonna throw 'em out.
Throw 'em out, Joe.
Really?
Skin got hard, yeah.
Mike and I laugh.
Skin got hard?
I had a sandwich today.
He had a sandwich today. Isn't that nice?
I'm yelling to Mike in the other room. Mike laughs harder.
I'll tell you something, jerk—let me explain something to you. You put rotten food in the refrigerator and you end up breeding germs in the refrigerator. And the rest of the food in there, whether it's rotten or not, that goes rotten.
Mike—can you believe this?
Mike is laughing his nuts off now.
Ya hear this?
You understand that or is it too much for your mind to comprehend?
Mike, ya hear asshole talking?
Does your mind comprehend that?

Do you believe what you just heard? Mike? Do you hear what he just said? That animal? He called my food rotten!

That's right.

Ya hear that animal?

I call it as I see it. And I just threw it out. So it's rotten, that's what I see.

Ya hear that animal?

That's the way I see it.

Ya hear that animal talkin'?

My mother harps on the fact that my father never wanted to leave Syracuse, *never even wanted to take a vacation out of town, mind you.* And she's right.

Your father would give you the shirt off his back. But he's a pisser.

She's right, he's a pisser: easily riled; stubbornly attached to his values; a poor, even insecure socializer; impatient with anything but wood and, over time, food. My mother, on the other hand, is patient, cosmopolitan, given to discussion and nuance, a versatile, accomplished woman.

Maybe she grows bored with him over the years, after the war, living in the suburbs. And maybe she can't know this will happen, doesn't know *herself* that well. Can't.

Two small black & white pictures of my mother, one a shot of her poised on wood skis in the Alps circa 1943, sit atop my nightstand. Two small black & white pictures of my father, one dated 1935 of a sandy-haired and strangely self-assured scamp in white button-down shirt, hover behind me here, as I work at my desk.

After the divorce, my father's fractured romances pile up.

He never reads about finishing. He knows how to work with wood, but like so many craftspeople, he can't entirely articulate what he knows. He's a professional, but he's no good at PR, and can barely earn a living at this trade. I buy him a book about finishing, but he reads only a few pages.

That's a helluva book, Joe.

He's a terrible businessman, terrible with finances—always bad with money, always behind in bills.

He doesn't keep a neat house.

He knows English inside out—the kind you put on a cue ball. He knows how to play a parlay, how to throw a left hook.

Odd thing, perhaps, but my father has a knack for crossword puzzles—the hometown variety, in the local paper. Maybe it's his facility with languages spoken. Sometimes he, Mike, and I work the *Post-Standard* "jumble" together, trying to get the letters to add up.

Still, aside from the newspaper, and save for what he manages to glean from a copy of *The Joy of Cooking*—a gift we buy him for his fifty-sixth birthday, the very same *Joy* that now sits, pages pawed and smudged, on my pantry shelf—he reads very little about cooking, about anything. Sure, if the refrigerator is near empty—and there are stretches when it is—he can put together a great meal out of just a few ingredients. Food or wood, the final product is consonant with my father's fussing.

Now, you don't have to know how to read or write in order to season food to taste, or restore a sheen to second-hand lumber. But it's one way to become a better cook, a better finisher. My father—whether making something out of something or out of virtually nothing, his method is very much of the moment, and he can't really articulate it. This sort of talent rarely lands you a job.

If you ask my father, say, how much pepper to add, he motions with his hands, pretends to shake a pepper shaker into his palm—once, twice, just like my gramma.

So you *could* say my father missed his calling.

Or—as some of my more literate acquaintances have suggested—that as a father, as a husband, as a companion, as a furniture man, as a single head of household, as a homemaker, even as a cook, Joe Amato, Sr., is a failure.

But did he fail Mike and me? And what's a calling if not one's obligation to circumstances, outside and within?

I would write of my father what Charles Olson wrote of his: *What can be lost is the weather of a man.*

Even a casual observer would conclude that my father's is a dying trade. And he knows it.

I've been lucky, Joe—I like my line of work. If only these bastards would pay better, we'd be back on our feet in no time.

He clenches his fists, shakes his head. Takes a drag off his Parliament, a swig from his can of Carling. He's angry.

Because they won't pay better. Because they'll never pay better. Not even ten bucks an hour, under the table. And he knows it.

With the advent of stressed furniture, plastic finishes, urethanes and veneers and non-wood laminations of all kinds—the appearances of appearance—most folks are just not willing to pay a reasonable fee to restore a beauty they've never known, or forgotten. Or can't know. At least, the folks *he* knows—people who have less and less money for the basic necessities. Hanson sells the furniture, reaps the profits. And who can blame Hanson?

And my father just doesn't have the temperament to move in what seem to him more sophisticated circles, circles that prize teak and rosewood, or solid mahogany, or baby grands. He just can't establish his own clientele—

he's no entrepreneur, not by a long shot. Doesn't *want* to be, in truth. He just wants to work with wood.

On house calls to impeccably furnished homes to repair nicks or dents on the premises, he's almost too humble, doesn't know how to make small talk. He walks in respectfully, his left hand gripping by its single leather strap his fifty-sixty-pound finishing box—more like a portable chest—finished by him in a dark maple hue, now nicked and dented and browned with age. He unclasps the box, exposing its top section—cotton rags, camel-hair brushes of various sizes, and two rows of aniline stains, each chromatic powder shimmering through its quarter-sized clear plastic container. He pulls out the box's front panel insert, revealing small drawers cross-hatched into tiny compartments, each filled with specialty nails, screws, tacks; alongside of which, two cubbyholes for his knife stove, a small jar of solvent or stain, a sanding block; above, a drawer for putty knives and lacquer sticks, scraps of sandpaper and wood dowels; below, a larger drawer crammed with screwdrivers, files, a hammer, pliers.

And he gets right down to work, stepping quietly out to the car if he finds he needs another tool, the odd bit of material, a spray can—but never for a smoke. He never asks to use the customer's bathroom. It's almost as if he isn't there—but he's there all right, deep in concentration. At times like these his hands are rock steady.

The customer is satisfied, but not stroked. And if the customer should watch him while he works, my father's face flushes imperceptibly.

I know—I've been there, with him. I can see him tense up, I can see the veins in his neck surface, even if the customer can't. But the customer can sense it, and customers are always right.

So help me Christ, Joe—

So word spreads, as words are wont to.

Joe Amato? Helluva finisher. But a real pain in the ass to work with—a real hothead.

Hothead, right.

6.
Linkage

Real work was still in front of me. I had only been resting, gathering strength for an act of connection about to become clear.
                    —Richard Powers, *The Gold Bug Variations*

CENTRAL NEW YORKERS are given to exaggeration.

No shit. Everybody, the entire region.

Has to do with how bad things are. Bad weather, bad economy. And as they say, and as I've said, misery loves company. A kick in the ass.

And maybe it has something to do, too, with the high percentage of Italian Americans in this neck of the woods—among the highest in the nation in the middle years of the last century. When we can speak freely, we southern Mediterraneans like to lay it on the line.

Weather SUCKS—worst on the planet. Shit weather. Real shit weather. They have less snow in Anchorage, and Anchorage BLOWS. But Syracuse?—it's a great place to live, FUCK YOU.

This is the daily ritual: we let everyone within earshot know that worst has come to worst. So there's nothing to worry about, cowboy—this is as bad as it gets.

We enjoying ourselves? Tell me about it.

Just don't expect me to give a shit.

Check it out though: worst *never* comes to worst. I mean, things could always be worse. Deep down inside we all know it.

Me, I'm something of a poet by the time I start drinking—or so I think. At heart, and even today, something of a precious child, but fired some in the social kiln and learning, still learning when to hold on to change, and when not to. It's an attitude, then, more heart than mind but ever mindful, a myopic lad but no less for that a lad attuned to his field of vision. In particular, he's always looked forward to those first few flakes of white, the generous calm with which the earth accepts this harsh edict from the heavens above.

And if that's poetry, it's precious as hell. But why not.

Comes right around Halloween usually, followed by the first real snow of the season, hush of white on white. Sometimes Christmas is green, but if it's a rough winter, bout after bout of snow flurries, snow showers, and squalls will have become a chore by the New Year. Chore as in backache, assache. Rough as in Tell me about it. Just don't expect me to give a shit.

Let it snow, let it snow, let it snow—after all, there's nothing to be done about it. But this is not your picture-postcard Vermont, brother. Not your Holiday Inn, sister.

Wise up.

1967 Bel Air.

Two-speed Powerglide automatic, steering-column shift. Detent indicator on the steering column reads Park-Reverse-Neutral-Drive-Low. Couldn't get out of its own way with a Twenty Mule Team, but it runs. And runs.

The transmission mount is busted—you can hear the tranny smack up against the floorboards whenever you hit a bump. Because the transmission is moving around so much, the linkage keeps getting stuck, usually in Drive. Each time this happens I have to jack up the car, get underneath, free up the linkage.

But until I get around to it, the tranny is stuck in Drive. No Reverse. You can move the detent to Park to start it, but the car lurches forward, because it's still in Drive. So you start the car with one foot on the gas pedal, and one on the brakes.

Driving around like this is an exercise in spatial relationships. You can go forward, but it's best not to pull in anyplace where you'll eventually have to back up. In parking lots, you hope like hell somebody doesn't pull up in front of you.

After drinking too much at the Poorhouse North one snowy Friday night, and after wolfing down the obligatory frittata, toast, and three cups of coffee next door at Two Guys from Italy, I find myself at three in the morning half sober, and boxed in. I have to head back into the joint to ask whoever owns the Dodge truck to please back it the fuck out of my way.

On my way home, I'm getting more and more aggravated just thinking about the linkage. I figure I might as well fix the cocksucker now, so that I don't have to look forward to freezing my ass off in the morning. But either way, all I know—or all I care to know in this state—is that this car, this fucking piece of SHIT, is responsible once again for me freezing my fucking ASS off.

By the time I reach home, I'm furious. I slam the car to a halt at the end of the street, wedging it tight up against the snowbank on the passenger's side. Positioning it thusly against the snowbank will give me the room I need to get under it on a flat level surface. Ramming it thusly against the snowbank will satisfy my impulse to drive this fucking piece of SHIT off a steep fucking cliff.

I stumble out into the darkness, a biting wind hitting me right in the forehead. I trudge through the snow into the shed, where I've stored the full-ton floor jack I'm borrowing from Frank. I pick it up—it weighs fifty pounds if it weighs a pound—and stumble back over to the car. As I approach the Bel Air, I put all my strength and one loud grunt into heaving the jack at it. It thuds up against the driver's door and bounces off, flopping tits-up onto the road.

Now I'm ready to get down to work.

I run upstairs and get a screwdriver, a pair of pliers, and flashlight. I try to be as quiet as possible, not disturb my father, who's snoring away on the couch.

Back downstairs, I position the jack under the frame and jack up the car. I trust this jack, so I don't bother with jack-stands, or with the rail tie I have lying around the other side of the house. Besides, I'm in a hurry, it's fucking cold. I wriggle under, my back cold on the snow and ice, and place the flashlight under my jaw, angling it so that the beam shines on the linkage. I start to pick at the linkage with the screwdriver, but something is wrong. Getting my head as close up as possible, I can see that one of the linkage rods has snapped in two, just above the transmission.

The following morning, a bright cold December day, I jack up the fucking crate again, scooch underneath, and remove the threaded steel rod, now in two pieces. My father takes the piece over to a friend of his at a nearby garage, who welds the broken rod back together. I reinstall it. The job takes a couple of hours, from start to finish. In all, on the surface of it, one of my better-executed repair jobs.

Only one problem: when I reinstall the linkage, I neglect to check that the transmission detent reads Park when the transmission is actually *in* Park.

I get in the car to start it, check things out. As it starts, it simultaneously kicks into Reverse. I slam on the brakes.

FUCK.

Fuck it, it runs, I'm not crawling under there again. You just wait a moment, then shift it into Neutral.

* * *

It's snowing like hell. The state authorities have declared a snow emergency, the radio jocks are telling everyone to stay home. It's so bad out they've closed all the roads in Onondaga and neighboring counties, and discontinued all emergency traffic — no cops, no plows. You get on the road, you take your chances. "Blizzard conditions."

For Mike, Rick, and me, this is the ideal time to visit Stan. The roads will be empty, no interference from asshole drivers who don't know how to handle the snow. Stan is living in Morrisville, taking classes at the tech school there. We call it Mooville. It's about a thirty-five-mile drive a bit south and east, down Route 20. Beautiful country.

We'll be driving the Bel Air. It starts in Reverse.

We hop on 690 eastbound, exit at Route 5 — New York State Route 5. Head down 5 a spell, bear right at the turn-off to Manlius. Through Manlius on Route 92 — concurrent with New York State Route 173 through town — on our way to Cazenovia. At Cazenovia we pick up Route 20 — that's U.S. Route 20 — and it's a straight shot to Mooville. Smooth sailing the entire way. We spot a pick-up off the road, flipped over. Nobody inside.

Look at that stupid asshole.

By the time we get to Mooville, the winds are gusting to 30–40 miles an hour, it's snowing like hell, and the temperatures have just started their descent below zero Fahrenheit. We don't know exactly where Stan is, because we don't know the campus. And because he doesn't know we're coming to see him, it's an Easter Egg hunt in the snow. We drive over to the dorms, and trudge around through the snowbanks, bootless.

Fucking cold out here.

Yeah.

We eventually locate his dorm room, but he's not in. So it's over to the Cherry Valley Inn in town, to see if he's there, getting loaded. We're freezing our nuts off, but they card us at the door, demand to see our Sheriff's ID. That's the ID we get when we turn eighteen, old enough to drink. The ID that requires our thumbprint, the one we figure they send to the FBI. Without a Sheriff's ID, no way you're getting into most bars most counties in Central New York.

Once inside, my eyeglasses fogged over, but I spot Stan. Sure enough, he's half in the bag, like everyone else in the bar.

Heyyyy — you guys! What the fuck!

Hey Stan!

Let me buy you sacks of shit a drink!

We walk over to the bar together. Stan calls to the bartender.

Sammy, a round of Molsons for my buddies.

Comin' right up.
When we each have a beer in hand, Stan raises his beer in the air.
Nazdrovia!
We all drink, play some foosball. Get fucked-up. I share a bed that night with a woman I don't know. No sex, just body heat, fuck me.

The next morning, the snow has subsided. It's still a bit windy, and the radio jock reports that the local temperature has dipped to thirty-five below. We eat breakfast in the dorm cafeteria, heads pounding.
Better go start the car.
Still groggy, all three of us walk outside. The cold hits us hard. Stings.
FUCK!
We hear the familiar racket of someone goosing an engine, the whine of tires spinning uselessly at each surge. Some asshole in a Firebird has managed to get stuck at the exit of the parking lot, blocking the way out for everybody.
Asshole's managed to get snow piled up under the center of his car. Real pain in the ass to get at with a shovel. ASSHOLE. We spend a half hour or so digging him out.
I get into the Bel Air, cross my rapidly numbing fingers. I turn it over, it begins to start—but stalls out just as it kicks into Reverse.
Shit!
I get out, pop the hood. The battery has no water in it. I pop the radiator cap. There's a layer of ice over the top.
FUCK.
Stan grabs a friend of his with jumpers. First we have to move the car out of the parking space to get at it. I shift it into Neutral, and with the driver's door open and my left hand on the steering wheel, lean against the door jam and push it backward with my friends till I'm clear. We hook up the cables, wait a few minutes. Then I get in and turn it over. Starts right up, kicks into Reverse. I shift it back into Neutral, let it warm up.

Some cars are reliable, no matter how beat-up. Others are lemons. Something is wrong with them way down deep, something in that intricate mechanical synthesis that stubbornly refuses to synchronize itself with human use. I've always thought it had to do with a resonating combination of factory errors, but who knows? What's important is that most of the vehicles we own are beat-up, and reliable.
A reliable vehicle is defined as one that will take you where you want to go, when you want to go there, with minimal downtime. By this standard, our '67 Bel Air is reliable.

Another reliable vehicle is Mike's '74 Yamaha RD350. Two-stroke twin, five-speed gearbox. The bike is his high school graduation present, again from my mother.

Mike drives it everywhere, at all times. It becomes an all-weather vehicle—rain, snow, ice, whatever. Under the circumstances, it might have been more practical to ask for a car. Some of the guys own snowmobiles in addition to cars—makes sense in this climate. But Mike and I aren't really into sledding. Like everything else at 501 Raphael Ave., we make the best choices we're capable of making at the time, choices that fall someplace between outright need and sheer desire. And Mike and I have always been fascinated with bikes.

Weather even remotely permitting, we take the bike up to SU together, my father driving to work in the Bel Air. Mike drives, I hang on with my left arm and both legs, one backpack strapped over my shoulders, and another wedged under my right arm. Most days it's a brisk, fifteen-minute ride, our helmet visors fogging the entire way.

Within a few years of Mike first setting eyes on his bike, Suzuki has become the bike of choice in the circles in which we move. Suzuki and of course Harley, the latter lingering on as a sort of legendary commitment to the road. Honda and Kawasaki are also popular in the region, but not in the circles in which we move. The brother of a friend's friend owns a Suzuki dealership, and GS750s start cropping up everywhere we hang out. Most guys we know install a header, and performance claims seem to rise with decibel increase.

And like everything else in Central New York, motorcycle ownership quickly reverts to party loyalties—which loyalties are in fact somewhat variable, permit for a constellation of four-letter antagonisms. Like beer—we drink Molson, Old Vienna—or national bands—we listen to Skynyrd, the Stones—or local bands—we like Todd Hobin, and Forecast, whose lead does incredible Hendrix covers. So Mike and I are pro-helmets most of the time—FUCK YOU. So Mike and I like Yamaha, FUCK Suzuki. And the gesture is returned in kind. Whichever bike you swear by, a friendly rush of adrenaline, male bonding at its most primeval.

Some issues cut across the party lines, to be sure. We *all* bitch about the bars closing so early—used to be 3:00 am, now it's 2:00 am—like we bitch about Sheriff's IDs. But we do little to lodge any formal protests. Just bitch bitch bitch.

Years later, Mike purchases a Yamaha XS1100. And I buy a Yamaha RD400, the descendant of my brother's RD. But a friend of mine is selling a Harley-Davidson Café Racer, and I can't resist. Mike hangs onto to his RD350, even after our friend Sally accidentally backs into it on Dolores Terrace, crushing the oil tank.

So eventually, parked in the rear-front of 501 Raphael Ave. are no less than four bikes. One is usually being worked on, center stand atop a large painted plywood sign that Mike and Greg rip off from a construction project on Buckley Road. The sign reads, red letters on white:

BROTHERS HAUL ROAD
TRUCK ENTRANCE AND EXIT

Now, two-wheels are two-wheels—with the exception of Harley, that is, whose true believers rant and rave Made in the U.S. of A-llegiance as if their lives depended on it. Some Milwaukee lives *do* depend on it, job-wise, but the way we see it, on the road is on the road.

I'll hop on the Yamaha, pass a Harley on the road, wave. No wave back. I'll drive home, park the Yamaha, hop on the Harley, pass another hog on the road and wave, and receive a response in kind. Same day, different shit. But Harley is on a downswing, sales-wise, their prices exorbitant, their performance lacking, and Japanese bikes are here to stay.

So get used to it, pal. And I don't buy your Harley horseshit.

I'm right in the face of a guy, Jack, whom I've met through Matt. Matt worked with Mike at Xerox in Rochester—this is some years after my brother's stint at Xerox, around the time Matt is being born again. Matt himself used to own a Norton 750 Commando—one of Mike's and my favorite British bikes. Matt is in town with his wife for the day, visiting friends, so he's decided to give me a call.

Matt's friend Jack is your typical dyedinthewoolpainintheass Harley fanatic. Now I like Harleys, always have. Dick Italia owned an Electra Glide, his son an absolutely choice Sportster. But Harleys are not exactly what you'd call high tech, especially not during the seventies. So if you like them, you like them in spite of.

And Jack, this dyedinthewoolpainintheass is giving me the holy ancient Harley spiel about *panheads shovelheads knuckleheads Then Came Bronson ridin' easy at 140 mph while Jap bikes are rice burners ETC.* as we sit at a bar in a local tavern, half in the bag. And he's been giving me this spiel for an hour now. Plus he's an Elvis fan, he says, but evidently he hasn't seen *Roustabout*. (Elvis rides a Jap bike in that flick.) I just can't take this shit anymore.

Oh fuck you and your horseshit.

According to Matt, who phones me the next morning, Jack got so pissed off over my irreverence that he ranted on and on about what an asshole I'd been, drank himself into a stupor, walked out of the bar and stumbled into a ditch. *And he broke his fuckin' leg!* Matt is giggling as he tells me this, I'm laughing my nuts off. If I'd have told Jack what I didn't know then—that McQueen himself had done a Honda commercial—he might have broken *both* his fuckin' legs, that fuckhead.

Now me, I'm a fair-weather rider. But I have my moments.

One night, late, early fall, we decide to head over to the Slide, a new hotspot downtown, right off of North Salina Street. Though we tend not to hang out downtown, we know this neighborhood. My grampa and gramma live less than a mile away, down State. The Slide is a block away from Columbus Bakery, a few doors down from the Nettleton Shoe Shop. Up the street a bit is Thanos's, one of the area's best Greek-Italian import grocery stores, right around the corner from which is Café D'Italia, where they make the best lemon ice in town. And across the street is Tino's, where they bake great pizza.

The Slide's claim to fame is a second-floor spiral slide that drops down into the middle of the dance floor. Mike hops on his GS1100, I hop on my hog. Frank rides with Mike.

We get over to the Slide, pull in right in front, parking at a slant alongside the row of bikes already there. It's still warm—we're wearing T-shirts, jeans, sneaks—a little breezy, a few red and orange leaves waft around the street. I try to keep the Harley at a safe distance from the other bikes. We unstrap our helmets and walk up to the door, where an unusually large bouncer sitting on a barstool asks us for proof. We show the tough guy our Sheriff's IDs, and walk in.

It's your typical disco scene: dark, lights whirling off of ceiling fixtures and reflectors, music blaring, bodies in motion on the dance floor, some actually keeping time with the beat, some trying to. Some of the music is disco, some is rock. The crowd is mostly white. We walk up to the bar, Mike orders three Molsons.

No games in this place, so we stand at the bar, sipping our beers, watching.

Geta loada that chick.
Yeah.
Jesus.
Fucking-A nice ass.
Yeah, nice everything.
I bet her shit don't stink.

I bet it *don't*.

I bet she *thinks* it don't.

A few minutes go by, and suddenly we hear a commotion at the door.

Look at those dickheads.

A fight has broken out, and two bouncers rush over to break it up. They manage to shove the two dickheads out the door. The three of us instinctively set our beers on the bar and walk outside. The fight is taking place right in front of our bikes. You knock one over, you knock three or four over. So we stand outside, keeping an eye on things.

Mike walks over to his bike, unstraps the Frisbee he's attached with bungee cords to the back of his short sissy bar. We toss the Frisbee around in the street, out in front of the bar, for ten or fifteen minutes. Nobody seems to mind.

After, we lean up against the outside wall of the bar, catching our breath. A guy pulls up close to my bike. Real close. He gets off, his foot almost hitting my gas tank.

Hey!

Yeah? What?

He walks up to me.

Not so fucking close!

As I say this I zip the Frisbee at his bike, bouncing it off his gas tank.

OK, OK.

Surprisingly he doesn't make a fuss, gets back on his bike and pushes it forward and backward till he's a few more feet away.

Thanks pal.

We head back into the bar and get good and shitfaced.

I have to take a piss. When I walk into the restroom—a half inch of water on the floor, and it stinks—the two urinals are taken, a white guy and a black guy standing alongside each other, both wearing leather jackets. *You must be sweating your asses off*, I think. I walk past both and step into the stall—better anyway, because it always takes me some time to get to pissing in company. The toilet is backed-up, full of shit, but I piss into it.

I hear one guy—the white guy two urinals over—grab the handle to flush the urinal. Then I hear him step over to the sink, and I hear the leather of his jacket squeak as he turns.

Hey—we don't wanna see your kind around here.

What's that?

I said we don't wanna see your kind around here. You have your part of town—you understand what I'm saying?

I hear the other guy flush.

Yeah.

We don't want any trouble. You just keep your people where they belong, OK?

Yeah.

I hear the men's room door swinging open and close, the noise from the bar filling the restroom for a moment. Then I hear the other guy walk out, door swinging open again, more noise. I zip up, walk back over to Mike and Frank.

We get set to leave, finishing our beers and picking our helmets up off the bar. As we walk outside, the fight has resumed, both combatants going ape-shit in the middle of the sidewalk.

Don't these guys know when to go home?

Just like that, one of the guys falls backward, smacking his head up against my license plate bracket.

Christ!

I run over to the Harley, wheel it away. I look at the bracket, and it appears in this light to be tilted in a little. So I grab a hold of it and, gently and as close as I can eyeball it, bend it back to its original position.

We get on our bikes, start them up. My header drowns out just about everything with its low, throaty rumble. I'm drunk on my ass, like Mike and Frank—and what's more, we all three *know* we're drunk on our asses.

As we pull out onto Salina Street, Mike gooses it, and I follow. I can see Frank's body lurch a bit backward, and can see him crouch and hold on tight. Mike gets up to 85 mph fast, and I get up there in short order. We're driving on the main drag through the north side of Syracuse, at two in the morning. As we near intersections, traffic light green *or* red, we're downshifting and slowing to 70, then punching it full-throttle till we're back up to 85.

We hit Park Street and we're cruising, maybe 90–95. As we hit the curve onto Onondaga Lake Parkway, we open it up. Mike begins to pull away, I twist the throttle all the way and leave it there. I can barely see a thing in my mirror because of hog vibration. The speedometer reads 110 and climbing. We hit the curve under the bridge and I can just begin to make out a howl, coming from my rear tire. I keep the throttle wide open, glancing quickly at the lights on the other side of the perfectly still lake, how they reflect to form a mirror image.

We slow right down as we near the village. We drive carefully past Heid's, eyes peeled for Liverpool town clowns. As we pull up to Two Guys from Italy, all three of us have smiles plastered on our drunken faces.

Hit 120.

Yeah, I lost you at around 110. But I'm hearing this noise in the rear.

All three of us shuffle around to the back of my bike. No license plate. I look up, above the rear tire, underneath the fiberglass rear fender. The bracket, plate and all, is smashed up between the tire and the fender—I didn't bend it back out enough, the tire catching it and jamming it upward.

You gotta be shittin' me.

Lucky you didn't get a fuckin' blowout.

Yeah. Fuck.

Fuckin-A. Lucked out.

We enter the restaurant just in time to catch the first fight of the early morning hours erupt. This one ends up clearing the whole joint out into the parking lot. Luigi, the owner, calls the cops, is standing in the doorway of his restaurant shouting and swearing, half in English, half in Italian.

It's a farce, and the three of us get a kick out of it. About the same kick we get out of watching The Three Stooges.

The next morning, bright and early, I disassemble the entire rear fender. Some job. Mike handles the wiring, my father and I repair the shattered fiberglass. I decide it's time to sell the hog.

Most Fridays and Saturdays, and some Wednesdays, we hang out at the Poorhouse North. The Poorhouse has a small outdoor patio. We'll get a case of Old V shorties on Friday nights, three for a buck. Bring it out on the patio, covered with ice cubes, and have at.

The crowd is white, and hardass, or would-be. They're not armed, but they act like it. Foosball or pool is always one wrong move away from a brawl.

The Poorhouse usually has a band on Friday nights. The guys stand around, stiff, tapping their beer bottles on their thighs, possibly tapping a foot or bending a knee in time with the music. The chicks, as we call them, cast glances here and there. Once in a while an ambitious couple walks to the small open area in front of the band and begins dancing. Usually the guy can't dance a lick, and the chick is showing off her stockings. Usually the rest of us roll our eyes at the guy's nerve. But we're envious, most of us.

One Friday night Julie's brother-in-law Derek and his band play at the Poorhouse. They're an easy listening group—do some toned-down rock, some rhythm and blues, and some pretty good versions of the newer Doobies. Derek plays keyboards and sings. The lead is a black guy, with a solid voice.

Diane walks up to me in the middle of one number. Diane is good people, has a heart—but like most of the women I meet out, comes with a tough exterior. Rough around the edges, and then some. Diane is Bill's wife, has been for years now. They argue a lot, and in public. Once in a while they bring their son with them into the Poorhouse. They know the owner.

Hey Joe.
Hi Diane. How ya doin'?
Pretty good. Wanna meet my friend Nancy?
Sure.

Diane introduces me to an attractive brunette. I've never seen her in the bar before.

So how d'ya like it here?
Pretty good, except for that nigger up there.
You don't like the music?
I can't stand niggers.
I think he's a good singer.
Can't stand 'em.

Lots of the guys at the Poorhouse drive bikes—a couple of choppers, but mostly those GS750s, fuck 'em, an occasional Yamaha like my brother's, and maybe a Honda or two. My brother pulls some incredible wheel-stands in the parking lot, front wheel hanging in mid-air for seconds on end as he coordinates throttle-twist with clutch-release. Even our buddy Nick Wilson, a real wild man, is impressed.

The platitude rings true—motorcycles are accidents waiting to happen. For us, that's part of their attraction, and not unlike our lives in this regard. So you have to have the right temperament. And even then.

In the circles we move, those of us who ride bikes all manage to dump them at least once. Mike, Dan, Stan, me—we've all been lucky when we've hit the pavement. Road rash, a few stitches, wounded pride. I drop mine once right in front of a crowd down at the lake, end up with a back full of scabs for the summer. Mike always says it's best to pick a cow pasture when you want to get squirrelly. But boys will be boys.

And Mike knows this as well as anyone. Headed north on Buckley Road one afternoon shortly after getting his motorcycle license, he notices the car in front of him swerve slightly to the left approaching an intersection, its left taillight blinking, indicating a left turn. Mike instinctively leans his bike to pass the car on the right— only, the car *turns right*. Wrong instinct. Mike ends up colliding into the side of the car, the impact not a quarter-mile up the road from where we used to pick strawberries as kids. Turns out the car's right brake light is out, and the car's driver is one of those assholes that tend to round out their corners.

No tickets are issued. But you see, you need to read the signs aright if you wish to cultivate the right instinct. In Mike's case it all comes down to some out-of-pocket expenses for repairs to his Yamaha—the owner of the rust bucket he rams cares not a damn about the dents and scratches—a badly bruised left elbow, and a clean half-inch hole in the front of his left shin (I helped the emergency room physician with the stitches)—in all, a lesson learned the hard way, and an arthroscopy years later to remove bone chips from his elbow.

Nick Wilson and his GS750 constitute a fatal accident that finally does happen. It may be an exaggeration to say that Nick is a wild man, that his death, while surely not in the cards, is not entirely out of character.

But anybody who knows him will tell you: Nick *is* a wild man, a real character—exuberant, full of life. He's not among our closest friends, but we're tight. It's through him that his circle of friends becomes, until his death and for a short time after, ours. And if you know how to read the signs aright, you can see it coming—all of it.

Nick is a prankster. Always sneaking up to tackle you, wrestle you to the ground. My father tells me that if you play around, you're bound to get hurt. Nick has never learned *not* to.

We first meet at the Gin Mill, up on Route 57. It's north of the village, in town of Clay. The Gin Mill has become our premier foosball bar. It's two-bars-in-one—half plays rock, the other half disco. The tables are in the rock half. You hold a table for three hours at the Gin Mill, you're kickin' ass.

I've put my quarter up, and I'm standing next to the table, waiting. A guy wearing wire-rims, fit looking, my size, walks up and stands next to me. There are five quarters on the table.

That's my quarter next.

I turn and look at him.

No—it's not.

Yes it is.

I've been standing here since I put that quarter up. It's my game next.

Yeah?

Yeah.

Our exchange has been easy-going, with little hint of tension. He looks at me, sizes me up. A mischievous look crosses his face. It's difficult not to like the way he handles himself, even when he's in the wrong.

That's your quarter?

Yeah, that's my quarter.

OK. You seem like a good guy, like you're telling the truth. Nick Wilson.

He turns slightly and reaches his hand around, offering it to me to shake. I do.

Joe Amato.

Joe *Amato*? You Mike's brother?

Yeah.

Cool—I know your brother. He's a good guy.

Yeah.

Yeah, has a great pull shot.

Just then a fight breaks out. Wilson and I shift our positions a bit as the commotion moves past us.

Check it out.

Yeah.

A fight breaks out at the Gin Mill every half hour. On a given night, the place employs anywhere from six to eight bouncers. One is a giant named Bigfoot, who must weigh five hundred pounds. Next biggest is Tiny, who weighs in at maybe three hundred. A guy named Hugh owns the place. Hugh is around our age. Nick knows Hugh from someplace or other and doesn't like him.

Sometimes, when Nick gets drunk and rowdy, they kick him out of the bar. It usually takes a big bouncer like Tiny to pick up Nick and muscle him out the door, because Nick can bend his athletic body this way and that while grabbing and holding onto every doorway and fixture around. Once outside, Nick stands in the parking lot, shouting out to Hugh to show his face, challenging him to come out and fight him. Hugh never shows.

Smart businessman.

And then it happens. Late one night, after a party at Nick's. Everyone drinking, playing pitch, Nick shooting the moon too often, as usual. He's moved into an apartment out in Lafayette, a few miles down off of Route 20, west of Route 81—out in the boondocks, pretty farm country south of the reservation. We all leave the party late, or early. Next morning, I get a phone call from Julie.

Julie had been seeing Nick for a few months, about six months prior. It was a difficult time for me—Julie and I were on the off. Or off and on, as we had been for some years. And I liked Nick, even though he was always on the make.

Anyway, when Julie called, Mike and I made some phone calls. Turns out that, after the party, Nick had left the apartment on his bike, half in the bag. Had run into some folks who'd run out of gas on 20. So he grabs their gas can, speeds off to fill it up, without his helmet on—typical hero-to-the-rescue antics. Takes that long steep curve on 20, just west of 81, way too fast. The same curve he tells Mike he almost lost it on once.

Dumps it, sliding headfirst under the guardrail.

And so a group of young men and women from Liverpool are called upon to attend a funeral, make like character witnesses. But the character in question is nowhere in sight—it's a closed casket.

I thought Wilson was gonna pop out from behind the church, tell us all it was just a gag.

Sal, Nick's closest friend, is trying to find words for it, his froggy voice straining. Bill is stunned, and Chris, his brother, is choked-up. Later that day, Mike, Sal, Julie, Vic, Bill, Chris, and I are sitting on Rick's front lawn. For a brief moment, we all constitute a link, a link from there to here—from that peaceful, dreadful silence to the noise of traffic. Nick's final prank is for keeps.

For keeps.

Sad thing is, you get used to it.

After Nick's death, the two sets of friends drift apart. Most of us are college-bound, or college grads. Most of them are high school grads, working construction now or factory jobs. I used to think that such differences were entirely reconcilable. Now I know better.

The ups and downs accumulate, like snow, like Februaries. "The most serious charge which can be brought against New England," Joseph Krutch once observed of that austere month. He should have spent more time in Central New York, where the easternmost Great Lake precipitates the sky.

Purification through sacrifice: Syracuse audiences once served as the upstate tryout for Broadway-bound plays. Dog town, though it's since become home to a fine regional theater. Hard rock, southern rock, acid rock, metal—they've all found a place with this caustic crowd, many of us learning early in life to talk the talk, and salt away the tears.

We the people, we the people who are purged of our sentiment, trudging ineluctably forward, forward through all reversals of fame and famine and fortune, stuck in our reliable and unreliable natures, built and building this way and that, bluffing our way through.

Story of my life. And I mean, things—

Part II
Rebuilding

7.

Salt City

> Not only did New York have a Venice, it had a Liverpool, the town being named so that Onondaga salt could be shipped around the United States with that trusted old brand name "Liverpool salt."
> —Mark Kurlansky, *Salt: A World History*

I RECALL MY FATHER pissing in the lake on a sunny afternoon, after a hotdog from Heid's. A park cop drives by, spots him, and pulls up. The cop gets out, asks him what he's doing.

What the FUCK does it look like I'm doing? I'm making my contribution.

My father grabs the half a hotdog sitting on the hood of his car and whips it into the lake. When he gets mad—irritated—like this, his nostrils flare. The cop walks away.

Before the Civil War, the salt springs bordering Onondaga Lake, town of Salina, produce more than half of all the salt used in the states. Solar salt fields line the lake, with a towpath canal for transport, a few miles as the crow flies to the Erie Canal. The four-foot-deep canal runs smack through the center of Syracuse, where it intersects with the north-south Oswego Canal, and the bridge-scaped town, as Kurlansky reports, enjoys some notoriety as the "American Venice" of the mid-nineteenth century.

As the century winds down, the salt reserves dry up. Clinton Square—named after the governor of New York most responsible for the building of the Erie Canal, DeWitt Clinton—now sits atop the old canal intersection, and at Christmastime we drive downtown to ogle the large decorated tree that serves as the centerpiece of the square.

A popular amusement park attracts local residents to the north shore of the lake toward the end of the salt century, near the area now called Long

Branch Park. Where we party on occasion now—folk concerts, Oktoberfest—in the small outdoor amphitheater.

My father, like my grampa, recalls swimming in the salty springs that remain. Onondaga Lake Parkway was just a dirt road then, the canal still a feature of downtown Syracuse.

The brine would keep you afloat, head above water.

My grampa Rosario works for the Onondaga County Park District into his early eighties. He tends the flower gardens. Everybody calls him Roy. Even my father, most of the time. On hot days when we visit him and Gramma in their small upstairs flat, my father wets his hands with cold water, pats his father's few grey hairs back.

The lake emits a noxious odor when it rains. During the early years of the city's infrastructure development, storm and sewer systems are permitted to flow together. So for decades, the wastewater treatment plant at the south end of the lake has bypassed—*must* bypass—raw sewage during a downpour, the sewage flooding directly into the lake untreated. Throughout the early and middle decades of the twentieth century, pollutants from area industries go unchecked, as well. In addition to the sewage, two severe hot spots develop in the lake, where mercury and numerous solvents have accumulated over time. The mercury and solvents are by-products of major employers such as Crucible Steel, where Steve's father works, and especially Allied Chemical—the famed Solvay Process, a soda ash production process using ammonia and brine, on the west shore of the lake, near the fairgrounds. The lake's relatively rapid flush time—four complete changeovers of lake volume annually—can't offset the solvent and sewage loads.

Jobs are jobs.

As kids, we learn not to touch the water. Tiny shells line the shore. A former Jesuit mission outpost, Ste. Marie de Ganentaha, attracts visitors to the east shore. We call it the French Fort. We don't visit often. We don't know much about the Jesuits, or the Iroquois, or Samuel de Champlain. Or the Onondaga Indian Reservation, south of the city's south side, recipient of nine tons of salt each year from New York State—partial payment for lands taken under treaty. Or so claims the state.

Watergate is wrapping up.
What do you think of that, Joe? That dirty bastard.
He's smiling, I'm laughing.
Go on, get the fuck out of there.
We're in the living room, huddled around the tube. Mike is laughing too.

We're, shall we say, appeased. What next? It's been a couple of years since our last round of public assistance, but—

—We're broke, still. My scholarships are paying tuition, the leftover three hundred I give to my father to help out with the bills. I apply for a two-thousand-dollar student loan, put some real effort into the loan application, explaining as honestly as I'm able how the money will figure into our family finances. It's a long shot, but we're hard up.

The bank calls—they've approved the loan. My father expresses his gratitude—to me. This relieves some pressure, especially now that Mike is planning to attend SU in the fall, double majoring like me in math and mechanical engineering.

They say the best jobs are inside jobs. In May, Helen puts us on to this company she works for, Salt City Auxiliary. Part of the VFW. Or so they say.

Salt City Auxiliary sells patriotic junk, by phone. They pay a fixed fee to the VFW to use their name. Helen works one of the phones. And they need delivery boys. We've got wheels, so we bite.

A buck per delivery, provided the customer actually pays for the merchandise.

Mike and I pull up to this four-five bedroom home in Manlius. Nice lawn, double-car garage, new shitbox sitting in the driveway. I get out and open the trunk. Mike pulls out the delivery notice, and tells me which small red-white-blue box, Made in Taiwan, to pick out of the larger boxes. Price is ten bucks. As we walk up the driveway, the front door opens.

Good afternoon, sir. We have a delivery you ordered from Salt City Auxiliary? It's ten dollars.

He's staring directly at me. Well dressed, maybe forty. I can't help looking over his shoulder, into the sunken living room in white shag, the solid mahogany hutch, the kitchen with built-in bar.

The guy looks down at the boxed bauble I'm holding in my hand.

We didn't buy that piece of shit.

He shuts the door.

We pull up to a poor neighborhood on the south side. The lawns are small, the houses are small. Like where we live. Cheap postwar construction. I ring the doorbell of what looks to be a two-bedroom Cape Cod. A small old black lady opens the door. I pitch the merchandise, she lets me in. She walks me into her living room, asks me, in a soft voice, to wait a moment. Though the day is sunny, the living room curtains are drawn, the room bathed in shadows.

To my left, there's a shrine of sorts. I shift a bit closer, peer at it. Two small American flags, crisscrossed. Two lamps illuminate a number of old black and white photos. The centerpiece consists of three portraits of three men, all in uniform. One is a good deal older than the other two.

She returns with the money.

On a good day, we net maybe thirty bucks, minus gas.

I run into Keith, my friend from high school, at the Cinema East. He tells me about a non-union construction job, building a new apartment complex up in Bayberry. I decide to give it a shot.

My father drives me up, waits in the car. When I walk over to the site, I can see a swarthy-looking man barking orders. He's wearing a white hard hat. I summon up the male in me, try to be forceful.

Hi, I understand you have openings for laborers?

Only for those who can pull their own.

Well I can.

OK. Tomorrow morning 7 am. Don't be late.

And I report to you?

Who the fuck else? Yeah. Fran Dell.

I offer my hand, we shake.

I turn and walk back to the car.

Who's that, Joe?—the boss?

Yeah, Fran Dell.

Fucking wop, changed his name. Probably a Dellapino or a DeLaurio.

The next morning, first thing Fran has me do is carry forty buckets of paint a hundred yards across the construction site and into a near-finished apartment. As I lift the last bucket, I'm nearly out of breath, my shoulders are straining, and my ankles ache.

Hey, take it easy there, boy. Wanna save yourself for later.

He seems OK. But something about him puts me off.

The next day, digging a drainage ditch with three other guys. It's hot as hell. After an hour of digging, I've developed a bad blister, in the middle of my left palm, right where the shovel handle sets. Suddenly, we all hear some shouting. We stop digging.

GET THE FUCK OFF THIS SITE!

But Fran, listen—

Hit the road, pal, hit the FUCKING ROAD. I don't have time for FUCK-UPS like you.

The carpenter tags along, pleading, a guy around thirty. But Fran ignores him. Ashamed, we're all ignoring him as well. Or trying to.

Fran registers the momentary work stoppage.

What the FUCK is everybody looking at—GET TO WORK!

The carpenter turns slowly, picks up his few tools, and walks off the site.

When we stop for break at ten o'clock, everybody heads over to the snack truck. Fran walks by.

I should tell this fucking PRICK not to stop by. Fucking long-haired hippie FAGGOTS. Fucking lazyass PRICKS.

My hair is just above shoulder length, hanging all over the place. Like a number of the guys.

I figure my days are numbered. The pay is OK, three bucks an hour, but Fran harasses us every chance he gets. If the concrete truck arrives at lunch, he works us through lunch. My father isn't real happy when I tell him about this.

He's a goddamn slave driver, Joe, like alla these wops who get a little power. Fucking guinea bastard. Probably voted for Nixon.

One day three weeks later, after heavy rains. The construction site now a mud-fest. Fran pulls me from some inside work and walks me out into the mud. He uses his arms to indicate an area of about five feet by five feet, and points to the center.

There's a water valve down in this location, maybe four-five feet deep. I want you to dig down, here, and clear it out around the valve.

He points to the center of the area, walks away. I start to dig.

You can't dig straight down—it's all mud, and every shovelful fills with water. You need first to create some room to dig—to be able to push your shovel head in on a flat and pull it out, without having to twist and pry it vertically, and underwater, to boot. So I start a few feet away, and gradually trench down toward the location Fran has given me. Fran returns.

What the FUCK do you think you're doing?

I'm digging a hole.

Who the FUCK taught you to dig like that? I said over here—the valve's down here, not way the FUCK over there.

You can't dig straight down.

Gimme that fucking shovel.

I try to hand him the fucking shovel but he grabs it out of my hands. He pokes the shovel down into the mud, and leans all of his two hundred pounds into it. It barely budges, and as he's struggling to pull it out, the shovel head buried almost instantly in a foot of water, I smile, looking right at him.

My grandfather can dig better than you.

Fran looks up, puzzled at first.

Is that a fact?

Yeah.

Is that a FUCKING FACT?

Yeah, that's a fact. My grandfather can dig *better than you*.

Well then you can just hit the FUCKING road. Just get the fuck out of my sight and HIT THE FUCKING ROAD.

Fine. I'll be back for my paycheck in a week.

I'm still smiling.

A week later I walk up to Fran, sitting atop a dozer. When he sees me, he looks down with what seems to me a moment of regret. But it passes. He reaches into his jacket pocket, and hands me the envelope with my paycheck.

Thanks.

He nods.

That same summer. I can't get a full-time job to save my life. Just odd jobs. Moving organs and pianos for a day. Helping my uncle, Sam, junking.

Middle of the summer, Helen has switched jobs. She's now at Big Ben Chemical, a warehousing operation in Solvay.

They say they eat cats there. All my life I've wondered why they say this. Doubt I'll ever find out.

Helen works all day packaging dye, scooping it out of large drums and pouring it into small containers. The dye gets all over everything—clothing, skin. The stains don't wash off.

She tells my father that they have an opening in the warehouse area. Minimum wage, $2.50 an hour. Sounds good. Only three weeks left till classes start, but what the hell. My father is working down on Erie Boulevard at another finishing shop, making eight bucks an hour under the table. He can drop me off on his way to work.

The next day, I'm standing in a dingy office, in front of this short burly man, Clint. Clint tells me that they handle bags of salt. He says it's heavy work, and he asks me if I think I'm husky enough for the job. Like any young guy, I nod.

He leads me out into the warehouse area. The floors are wood, the entire place a musty odor, punctuated by whiffs of exhaust from the few tow motors that whiz by. Clint introduces me to Jeff.

Jeff is my age, a bit shorter. But leaner, more muscular. Like me, he wears glasses. Like me, his hair is long—longer than mine. He seems down-to-earth, even friendly. He's missing one front tooth, has chipped another.

C'mon. I'll show you what to do.

He leads me through the warehouse, across a threshold, into what looks to be a long, narrow room. Maybe forty feet long, ten feet wide. At one end of the room is a pile of large brown bags, neatly stacked. Each bag contains roughly one hundred pounds of salt—calcium chloride, the deicing agent that, in its liquid form, is used to treat the sand and rock salt mixture they throw on the streets winters.

Jeff pulls a pallet—what I learn later is a forty-eight-incher—out of a pile of pallets. He throws it on the floor, in front of the bags. Like our voices, the thud echoes as though the room is hollow. Then he shows me how to grab and stack the bags. At first the weight seems severe. But after ten or fifteen minutes, I get used to the weight. I must.

As we finish stacking each pallet seven bags high, or twenty-eight bags total, a tow motor enters the room. The driver, a stocky bearded man, aligns the forks with the pallet carefully and quickly, and carries it off. Sometimes he arrives just a bit early. So he sits on the tow motor, goosing the engine, while Jeff and I hurry to finish. Jeff grimaces.

Fuckin' straw boss asshole.

I smile.

I hear you—but he can too.

I don't give a shit.

About a half hour passes, and we finish stacking the bags. The room is empty now.

Ten o'clock, time for a fuckin' break, let's go.

Good thing, my hands are getting raw.

Jeff smiles. He takes me over to another part of the warehouse, where I meet four other men. Otis, a bit older and taller than me, heavy set, thickly muscled. He makes Jeff look small. Eddie, lean, worn, in his thirties. Wimbleton, a bit older too, a strapping six-footer who says he's there to condition himself for football in the fall. And another guy, who evidently used to work there, a friend of Eddie's. Otis, Eddie, and Eddie's friend are black; Jeff, Wimbleton, and I are white. Otis, Jeff, Wimbleton, and myself are the laborers. Everybody is a laborer at times, though Eddie mostly drives tow motors these days. Otis and Jeff sometimes drive tow motors, but only when there's no lifting to do.

We're eating donuts drinking coffee everybody seems friendly. I'm enjoying myself. Red, the bearded foreman, is nowhere in sight. Nobody likes Red, who's white. Clint walks by—everybody likes Clint. Clint is white, too.

C'mon guys. Time to get back to work.

Otis is smiling at me.

You think you gonna last?

What do you mean?

Last college boy we had lasted one day, that's it.

He's still smiling.

Something about this bothers me, I can't say what. But it's clear we like each other.

After about a half hour, the group breaks up. Time to get back to work.

Jeff leads me over to another threshold area. But this time the entrance is closed, and I can see clearly now that the threshold is in fact a loading dock. And I suddenly understand that the little room I was in was the inside of a rail car. This time, the rail car doors are banded shut.

Jeff grabs a large clipper off of a stack of bags near the entrance. He snaps the metal banding, and pulls open the doors. As he opens the doors, I can see that the entire car is chock-full of hundred-pound bags of salt.

It's a wall of salt bags. I walk up and peer in, looking left and right. A forty-foot-long rail car filled to my height with hundred pound bags of salt.

Don't worry—we have all day to unload this one.

All *day*? You mean we've got to unload this car *today*? *Every fucking bag*?

Jeff nods.

All one thousand of 'em. Here, I'll start this one.

He grabs a pallet, and places it over the gap between the car entrance and the loading dock. As he's working, he explains the obvious—that there's no room in the car yet to place the tow motor ramp, the same ramp over which we apparently walked earlier. Something else I just didn't see. Then he spreads his feet apart, one foot on the ledge of the car, the other two feet away, on the concrete ledge of the loading dock. And he stretches up, grabbing each hundred-pound bag and guiding its weight so as to form a perfect stacking pattern on the pallet. UMP. The waist-high bags almost slide on. UMP. UMP. But when he gets to the bags toward the bottom, he's forced to pick them up and pull them into place. UG-UMP. They're warm, the bottom bags, some soaked with moisture the salt has absorbed, the salt itself melting in the process and leaking from the bags to produce a strong brine solution that covers the rail car floor. UG-UMP. UG-UMP.

It takes us two hours to unload the entire car. My hands, my fingertips, have begun to blister. And the salt has gotten into the wounds.

The job is now clear: two men, two cars a day, two to three hours a car. Sometimes a forty-footer, or a thousand bags—sometimes a fifty-footer, or twelve hundred bags. A thousand bags minimum, about three-dozen forty-eight-inch pallets. That's a hundred and forty *feet* of pallets, butted together, nearly head-high. That's two such rows, in four hours. That's one row a man. That's four bags a minute a man, on average. Or a bag every fifteen seconds.

Steady, for four hours straight. That's forty-eight tons a man in a half a day, on average.

At $2.50 an hour. Depending on how early you finish, fuck-off time to spare. Otis and Wimbleton unload two cars *each* a day, by themselves. Jeff stands in awe of Otis.

Otis's pallets are neat, man. Wimbleton just heaps 'em the fuck on.

At this point, neatness is not what most impresses me. But I'll learn that, even here, appearances matter.

That first night after work, my father gets off early and picks me up. He has to stop by my grandparents' on the way home. We park right in front on State Street, and walk upstairs.

Look at Joey's hands, Pa.

I show my grampa my blisters. He's hunched over from a lifetime of labor. He stares down, shakes his head slowly.

Wait minute.

He shuffles over to the porch, comes back with a pair of gloves.

Here. Before should use these.

Thanks Grampa.

He shakes his head again.

The next day, I appear with the pair of gloves, my fingertips raw, blistered. I'm to work with Otis today.

Otis and I walk over to a rail car. He opens the door, and we begin. It's a little after 7 am. We hit it off.

But my hands are hurting like hell. I try to pull the glove off, but my fingers stick to the inside. When I get the gloves off, my blisters are oozing. Both hands, middle fingers, forefingers and pinkies. Otis is concerned.

You ain't gonna work like that.

Watch me.

Rub some dirt in.

Huh?

Rub some dirt into the blisters. S'what I do.

I don't follow his advice. Too late for gloves, too late for advice. I work that day till my blisters pop and peel. I work till the pink skin underneath begins to bleed.

The ritual is so: I come to work from that day forward with gauze and dressing tape, and Otis helps me bandage up my hands. I take Otis's kindness at face value—he is genuinely concerned about my hands. At the same time, he's generally friendly, seems to want to get to know me, so I reciprocate his overtures.

When Otis is through with the wrapping, I look like Karloff's Imhotep come to life. But it kills the pain.

Otis lifting bags a sight to see. He smokes a cigarette usually, breathes in and out through his nose. When he gets going, the hundred-pound bags seem to fly up, into his arms, and fly out, onto the pallets. Flawless coordination, timing.

One day he and Jeff decide to have a contest. They'll pick up two bags at a time.

Jeff grabs two, and strains as he lifts them onto the pallet. Otis begins to huff as he lifts two. Then another two, then another two, then another two. UMP. UMP. UMP. UMP. Jeff stops, and Otis's upper arms expand to tree-limb size. Two bags fly into place with a heavy thud every six or seven seconds. UMP-UMP. UMP-UMP. Red smiles as he sits atop the tow motor. Jeff smiles, stopping. I smile. Eddy ducks his head in, smiling.

That Otis.

Turns out Otis hails from North Carolina. One hot August day, almost finished with a car, we shoot the shit a bit.

I used to work in a sawmill.

What was that like?

We'd cut the trees, then have to carry 'em down the hill.

By hand, walking?

Yeah. Was hard work.

And this isn't?

Not as hard.

Otis is married. I don't really get all the details. But his family has moved up to Syracuse from North Carolina. He lives someplace on the south side.

My hands are pretty much shot now, after three days of work. But I've grown used to the pain, and the other guys respect me.

First college boy who's stuck it out.

Jeff, Otis, and I horse around a lot. One day roughhousing with Otis, he grabs me by the inside of my leg and my waist, turning me sideways and holding me up in the air in front of him. I'm around a hundred forty pounds.

Put me the fuck down you fucker!

We're all laughing.

That Otis.

Salt City

* * *

One day Red is busting Jeff and me but good. Red is hustling, trying to push us to work faster. It's his mood, and his moods are plain mean. He keeps goosing the motor, his way of busting our balls. Eddie is in the shithouse, hungover. We don't have much choice but to try to keep up.

Let's show that fucker, that cocksucker.

We start to bristle. Before you know it, we're humping.

Let's go! Let's hump it!

Bag after bag after bag. We finish so fast we end up having to wait for Red, who's whipping his tow motor around like a madman. He's pissed now; Jeff and I are smiling. We're flipping him the bird behind his back, giving him the big double-fuck-you.

We keep it going. We lift and stack onto pallets five hundred and thirty hundred-pound bags in a half an hour. Neatly. That's one bag

every

seven

seconds

for thirty minutes non-stop. Apiece.

We showed that asslicking clit!

I'm laughing my ass off, exhausted. My back, shoulders, arms, hands, fingers are throbbing.

Jeff comes in one day, limping. He's fallen down a flight of stairs.

Joe, you've got to carry me today.

I do the best I can. Red pushes Jeff extra hard, knowing that he's injured.

The end of that first week I get paid for 39.9 hours. I've been docked for showing up five minutes late one day. My take-home is just under eighty bucks.

Most of the time I work with Otis. When we have no rail cars to unload, we're sent to the back of the warehouse. It's here we make pallets.

There are two large stacks of lumber. In between the stacks stands a metal table, its surface a template. Each man takes his place on either side. The job is, first, for each man to pick up the right number of slats, throw them onto the template, and arrange the pieces. Then each man grabs his large pneumatic stapler, hanging alongside the table from above, and as deftly as possible, staples the slats together.

CHOCK CHOCK CHOCK CHOCK CHOCK CHOCK CHOCK CHOCK.

Two hundred pallets a day. If you finish early, you fuck off. You can finish in four hours if you know what you're doing.

And Otis does. He shows me how to do it. At first the staple guns are heavy and awkward, but after a few hours, I don't feel the weight at all. I learn how to use the weight to my advantage.

So what's a white boy like you doing here? CHOCK.

I explain about college.

College huh? CHOCK CHOCK.

Yeah.

What you study? CHOCK CHOCK.

I explain about math—I always figure myself more the math major than the engineer.

I take math once. CHOCK CHOCK CHOCK CHOCK. Way back. But I had to leave school. CHOCK CHOCK. Hey—no, not that way, like this. CHOCK CHOCK.

He shows me how to staple a corner.

* * *

One day, the warehouse is empty save for Otis and me.
You look sweet. CHOCK CHOCK.
Yeah, sure. You like Foreman or Ali? CHOCK CHOCK.
You a sweet lookin' white boy, ooooh. CHOCK CHOCK.
He's smiling, half joking. Half not.
Lay off, will ya? Foreman or Ali?
George is a tough man, but—I like *you*. CHOCK CHOCK CHOCK CHOCK.
Yeah well I like you too. CHOCK CHOCK.
You do? CHOCK CHOCK CHOCK CHOCK.
Fuck you. CHOCK CHOCK CHOCK CHOCK.
Mmmmmm. CHOCK CHOCK CHOCK CHOCK.
C'mon will ya?—Foreman or Ali? CHOCK CHOCK.
Ooooh, sting like a bee. CHOCK CHOCK.
Will you knock it off? CHOCK CHOCK.
So it goes that day, and whenever we're left alone from that day forward. Otis never gets physical. And we don't talk about it after.

Toward the end of my second week at Big Ben, the tip of my left forefinger starts to swell. Within three days, my fingertip enlarges to half again its normal size. The lump hurts like hell, a dull ache.

My final day of work, three weeks to the day I started, the Friday before Labor Day weekend. Jeff cuts the banding off of the last rail car door. It's a fifty-footer. I have my back turned.

Uh-oh. Fuck!—it's tilted.

I turn to see all of the bags of salt shifted to the right. And instead of being piled on top of one another in vertical stacks, the bags are dovetailed, one atop the other.

The engineer jogged the fucker.

What this means for Jeff and me is that we need help. We get Otis. Each bag must be pulled, yanked out of its place, and lifted onto pallets. That's a tug of war lasting roughly one thousand one hundred and ninety-nine bags. Rough on your hands, your fingers. It takes the three of us three and a half hours. By the end of the day, when we say our goodbyes, I can't feel my forefinger.

On Saturday, the tip of my forefinger has begun to turn green. It's evening, and my friend Rick agrees to try to poke a hole in it with a needle heated over his stove. He does, and squeezes the finger. Fluid comes out, but not near enough. Julie tells me I should see a doctor. Julie's mother tells me I should see a doctor.

Lenny agrees to drive me to the medical center a mile down Buckley Road. My doctor is not in, he's off on vacation in Stockholm, so I have to see a guy named Paris.

The nurse is brusque. I figure she doesn't want to be working tonight, on a holiday weekend. She brings me to one of the examination rooms, and asks me to sit up on the examination table. Paris walks in, asks to see my finger. I hold out my hand, he places it in his. He takes one look at it, and shakes his head.

You stupid kid. You know what we'll have to do, don't you? I'll be back in a little while.

He leaves. I'm beginning to sweat. I shift my weight on the tissue paper and it tears. I stare at the diplomas on the wall. Paris is a former Army medical surgeon.

The nurse walks in and places a small bowl of something brown where I'm sitting.

Place your finger in this.

She leaves and returns with several items that she arranges on the nearby countertop: a scalpel, cotton swabs, an antiseptic of some sort. That's it.

After maybe fifteen minutes, Paris reappears with the nurse.

OK. Now lean back. This will hurt.

I lean back, my feet still dangling. Paris straightens my right arm, lays it flat, palm up. The nurse holds my arm down by grabbing a hold of my upper arm with both hands. Paris leans his left arm over my forearm, letting the weight of his body fall on it. He holds my right hand with his left hand. My arm and hand are immobilized. He presses hard.

I see his right hand reach for the scalpel.

Suddenly I feel a sharp pain. Paris squeezes. Then more pain. And more pain still. Paris moves slightly to readjust his grip on me. He squeezes hard. More pain. Somehow I feel his hand move up and down. I hear myself groaning, almost from a distance. I'm aware that I'm soaked with sweat. My eyelids are sweating, my eyes are burning. I look up at the nurse, she's looking down at my hand, wincing. I close my eyes, The pain lasts for three or four minutes, three or four hours. I can't tell.

That's it. Try to sit up now, slowly.

I sit up, in a daze. My finger is bandaged, I can't see it.

You OK?

I nod.

I'll be back in a minute. Just rest here.

Paris and the nurse leave. Then he returns.

I want you to start taking these pills immediately. And I want you back in here on Tuesday. You'll have to soak your finger in saltwater every day, fifteen minutes. It needs to drain. If this doesn't work, you may lose the tip of your finger. You understand?

I nod.

I walk out of the office, past the nurse, whose concerned look concerns me. Lenny is waiting.

Hey, you look like a ghost.

He drives me home. My ears are ringing.

When I walk upstairs, my father takes one look at me and asks me if I'm in pain.

Yes.

Did he prescribe any painkillers?

I don't know.

My father looks through the drugs.

No painkillers. Why that fucker. What's this guy's name?

Paris. Dr. Paris.

My father gets on the phone.

Dr. Paris? Listen, my boy Joey is here, and he's in pain. What the hell do you think you're doing? How the hell come you didn't prescribe painkillers for the boy?

A moment of silence while my father listens.

OK: listen, I want you to send a prescription over to Fay's. Now. OK?

My father leaves, comes back with the painkillers. They help.

When Paris sees me the following week, the wound has begun to scab. He has to break it open. I lean back, he squeezes.

Your father was kind of angry the other night, huh?

I groan.

Yeah, that's the way he is.

I understand. OK, keep soaking it. C'mon back in next week.

The next week I lean back, he squeezes.

As I'm healing, I try to understand what happened. The way I see it, I must have picked up a sliver over in the pallet area. The way I see it, I owe Paris a tip of a forefinger.

With the painkillers, the medical bill comes to sixty bucks. So for my efforts at Big Ben, I net a hundred eighty bucks for three weeks' work. Don't ask why we didn't try to claim it on workers' comp. The money goes into a new set of tires for the shitbox, to replace the old set worn out by my delivery job.

* * *

November. We're back on public assistance. Mike and I are both taking classes.

It's snowing hard, early season lake-effect squall. Helen calls—she can't get her car started. My father still at work, I tell her I'll pick her up.

Mike and I hop into the shitbox. Driving down Hiawatha Boulevard, skidding this way and that. The plows have been through once, but it's coming down hard. The sand & salt mixture flicks up off the tires into the wheel wells, a steady ticking sound. We pull up in the Big Ben parking lot, which they share with the Chinatown Outlet store. We walk in, I show Mike where I used to work while Helen gets ready to leave.

Hi Clint.

Clint shakes my hand.

This is Mike.

They shake.

Where's Jeff?

Oh, Jeff fucked hisself up but good, fell or somethin'. And just stopped coming in.

And Otis—where's Otis?

He put a staple through his hand one day. That was it for him, he never showed again. But I've got a coupla new guys. They're both used to this sort of work.

He motions over his shoulder, where two stocky black men in their thirties are leaning up against a pallet of salt, getting set to go home for the day.

Life is a lottery. You pay to play, the scarred prints you earn for your time tracing solvent conspiracies of industry and earth.

Five-six years later, I'm in a bar having a few beers. I get to talking with the guy next to me. He's maybe five years older. He says he used to work at Allied.

Oh yeah? I worked one summer for a few weeks right down the tracks from you guys.

Doing what?

Unloading rail cars full of salt.

No shit? We *loaded* those fuckers.

No shit?

Yeah. Bitch of a job. It'd take us a coupla days to get one boxcar loaded.

Huh? What do you mean?

A coupla days—we'd have six or seven guys gradually load 'em up, then send 'em down the fuckin' track.

We'd unload each car in two hours.

Huh? What the fuck you mean?

I mean two fucking hours. Two fucking hours, two fucking men. Two fucking cars a fucking day.

No fuckin' way.

No fucking way my fucking ass. Two fucking hours two fucking men two fucking cars a fucking day. We had guys could do it by themfuckingselves, a fucking man per.

You're shittin' me.

I shit you not. We had this one guy, Otis, black kid, could do two by hisfuckingself in a day.

Holy fuckin' Christ. Two fuckin' hours. Jesus fuckin' Christ. Minimum?

I nod. He rolls his eyes.

Jesus fuckin' Christ. We'd get paid ten-twelve bucks a fuckin' hour, plus overtime. Two fuckin' hours. Fuck me.

He shakes his head. He buys me a beer. I buy him a round.

Years later, there are still plenty of fish in Onondaga Lake. And warning signs posted to discourage unknowing fishermen from eating same. There's talk about developing the lakeside area, and bringing fishing back. To celebrate the fifth year of the $380 million cleanup effort, a family fishing derby is held—on what is yet one of the most polluted lakes in the nation. The year prior, Governor George Pataki had been joined by Governor Jesse Ventura to kick off the Onondaga Lake "Ultimate Fishing Challenge"—a promotional event intended to exploit the lake's potential for generating tourist dollars.

As to the pollution itself, public faith is put in diverting waste to faster downstream currents—the Seneca River, which leads ultimately into Lake Ontario. Maybe these currents will do a better job of aerating the sewage and pollutants, a better job of microbiological cleanup. Maybe.

On the east shore of the lake, a million bucks or more has since gone into restoring the mission. On the south shore, atop the ancient junkyard site, stands the large mall, adjacent to the wastewater treatment plant. And inside the mall, the restored carousel from the old amusement park goes round and round.

They say you can't step twice into the same river.

It's not right. It's just not right. Life is not a lottery.
Life is a lottery.

8.

Just Produce

A Meditation on Time & Materials, Past & Present

> But that which matters, that which insists, that which will last,
> that! o my people, where shall you find it, how, where, where shall
>   you listen
> when all is become billboards, when, all, even silence, is spray-
>   gunned?
>
> —Charles Olson, *The Maximus Poems*

LIKE MY FATHER, I've never been too good with money.
But I can never be my father's son.

My name? I drop the "Jr." to avoid redundancy. As I see it, I know who I am, and Joe Six-Pack, Jr., is more than a mouthful. But years later our credit ratings will still get confused—which makes it difficult for me to get credit. This despite the fact that my father, thanks to the nuns who drilled it into him, signs his name Joseph, not Joe—and not Giuseppe, the given name that becomes Joe by default on the sidewalks of Depression-era Syracuse. Joseph is the name he gives for his reissued birth certificate.

Sometimes it's best not to keep up appearances.

In spite of my father's talents, and with the exception of my rolltop, two dressers, a bookcase, and two antique piano stools he rescues from the trash heap, our furniture is a wreck. Reminds me of so many car mechanics I've known—so many of these guys drive crates. Why is that?

I do love tomatoes. Like my father.

Mike isn't too fond of them—raw. Though he's starting to eat them, now that he's turned forty, and finds them in his burritos. Cooked, they've always formed a staple of our diet, all three of us bred to simmering the berries stove-top, for hours on end.

I learn that I have bad gums, like my father—which he took to be bad teeth. Unlike my father at my age, I still have my teeth.

Sometimes it's best to keep up appearances.

Either way, appearances matter.

Still, tomatoes or no, teeth or no, I can never be my father's son.

How to measure the distance—between us, from here to there, and back.

For one, I don't have the lungs for it. For the fumes, the sawdust. I'm asthmatic—like Mike, but worse. We've both been through the allergy shot routine as kids, all seven years of it. Not much help, finally.

Like so many of my father's generation and background, most of my generation and background smoke, and smoke—but cigarettes *and* pot. Never mind that weed speeds up my already fast metabolism—I can't inhale it without some real hacking. So I find myself somewhat out of step with the times, and with the circumstances that helped to define the times.

As far as working with wood goes, Mike picks this up from my father. He's good with his hands, part of the reason why, even into early adulthood, he loves magic tricks. Orders all sorts of stuff from Vick Lawston's "House of a Thousand Mysteries" catalog. *I'm in business just for FUN*, Vick says in his catalog. As a kid, Mike likes surprise, sleight of hand—he likes to put on a show. I'm his assistant. We both have fun.

The divorce will sober him up some—permanently.

Me, I'm pretty good with a wrench. A decent cook. Maybe even a knack for colors, elementary renderings of landscape and still life.

But both of my parents have a habit of saying that it's what's upstairs that counts. From an early age, I understand what they mean. Maybe it's the look in their eyes as they say this. Or maybe it's that I'm well fed, live in a three-bedroom ranch, beneath infrequently blue suburban skies. Roof over my head. Whatever the reason, I manage fourth and fifth grades in a single year.

When we move into our second-floor flat at 501 Raphael Ave., my father brings that habit of saying along with him. Living upstairs, I understand even better.

An observation pertinent to the past century: a furniture finisher works with materials that rub off on him. And part of him, his very tissue, rubs off on these materials. This is his stock-in-trade. He works cognitively, as in all work, and his cognitive work must process the interaction of his senses with the stuff of three dimensions. Eyes and hands are his instruments, sight and touch the qualities he judges. Even when the piece itself is mass-produced, each bears his touch. Surrounded as he is by the odors of his trade, perhaps odor itself is a measurement taken—a sign of solvent dried, or mixture properly proportioned. You might, if you're lucky, receive a refinished piece bearing my father's invisible mark, expertly altered by him in some slight way. Lucky *and* unlucky, for such a piece will have been damaged, restored.

My father deals, that is to say, in the reparation of appearances, a beauty only and profoundly skin deep. And to the extent that each wood item is porous, like tissue, and singular—like each of us, like any *item*, so defined—it breathes his finishing touches, even as it resists his every effort.

He works through his body—as do I, as do all of those with bodies—to produce a crafted beauty. And it must be said: my father becomes, in the most literal sense imaginable, a corporeal tool of his trade, even as his trade becomes a tool of the corporation. For it is this same crafted beauty that will help to kill him.

Will help to kill him through the habitual circuits of life—inhalation, respiration, aspiration. As we—this technologically bound culture—will find more efficient ways to alter appearances, my father's masking tape masking the reality of things to come.

Whatever else, my father's lesson will not be lost on me: suspicious I am and will remain of the human and social costs of a craft that binds.

I work through touch too, as a (near)sighted individual—and some might say that I deal in appearances. To be precise, North American, middle-class appearances—at this moment, *now*. Typing is not a problem for me, anymore than is holding a pen. But like my father, I sweat the details. And what I work with, what resists my efforts more than material, are the virtually two-dimensional l-e-t-t-e-r-s *here*, on this page, and their relationship to the very stuff of thought. And if my computer mediates for me these mediating words, I am, signature to utterance, no less subject to their ingrained effects and affects. In some uncertain measure, then, I participate in a shared experience, even if the text before me, on this screen I'm busy scrolling, has not a trace more of my DNA than on those pages you're busy flipping, or on that screen *you're* busy scrolling, or in those sound waves you're busy processing.

People who lament the advent of things and words electronic often point to the tactile attributes of books as more befitting a creature with fingers. Ironically, however, with the exception of some vital artistic communities and their products, the production and proliferation of books has been given over largely to machines, the assembly line. Software has made even of typesetting a fraught trade. In the meantime, electronic media have opened up alternative vistas for authoring, communicating, publishing—various publics gone online, more and less purposefully—which has complicated the customary genres of writing-based products even as the business of publishing has become more and more economically tenuous, and merger-prone.

Is *this*—what you're busy reading, or listening to—a memoir, or a fiction, or an essay? It is certainly all of these, as any writer worth her salt will tell you. But what are the alternating effects of reading it as any one of these, and then as any other?

And at the level of material itself: the mass-produced book—*this* book, if it appears in your hands as a book and not an electronic copy—still requires for its production and distribution the social organization of machines and people. Today, however, one may assemble with one's *personal* computer— much as one might with a typewriter, or with a pencil—a uniquely handmade manuscript, virtually infinite variation possible within the regulated limits of software, printer design, paper type and size, and so forth. Or one may compose a uniquely handmade electronic text, with similar possibilities for variation. Or one may collaborate with others, across space and time zone and national boundary, to produce a somewhat different textual product, and an altered sense of writing process, authorship, and perhaps reading process.

Like the typewriter, or the pencil, or any device for making texts, the computer itself is hardly a simple machine, like the inclined plane; not quite a tool, like the wrench; not quite a machine tool, like the lathe. The computer, as part of a network of such machines, seems more and more to represent an *event*—telephone, radio, television, and library technologies intersecting with computational technologies—an event keyed to archiving information, communicating, computing, and producing texts. Composing, configuring, drawing, creating, solving, displaying, printing, copying, scanning, sorting, filtering, editing, corresponding, sharing, exchanging, swapping, discussing, arguing, addressing, confronting, alerting, assembling, simulating, arranging, combining, selecting, evaluating, saving, deleting, opening, closing, changing, searching, finding, buying, campaigning?— images, sounds, letters, numbers, graphics, data, feelings, beliefs, opinions, whims, ideas, moods, passions, wants, needs? As we find ourselves more and more interacting with, even *relying* on the range of possibilities it presents to us, the computer becomes a truly complex making machine, as complex surely as the assembly or packaging line, with its only occasionally intervening human attendant.

And whether machine or tool or machine tool, thus is process itself digitally enacted, reenacted—manually, in real time, as I type, or as I point and click. And thus does an analog world augur a digital rethinking of hand and mind, agent and actor, bone and silicon. "[T]o be as shaped as wood is when a man has had his hand to it," wrote Charles Olson of his projected experience of the word. Perhaps just so.

And all the while, the rich get richer, the poor get poorer, even as my middle-class wrists ache from *these* repetitive-stress patterns. I begin to wonder whether my new-age affliction is but a misbegotten variant of my old man's arthritis, different and same as we are, as we aspire to be ... mortal, iconoclastic, untutored in our failings.

A final consideration: aside from a couple of dressers he designs and builds for Mike and me, and a bookcase he makes upon request for my mother—all put together in the basement of our house at 112 South Dolores Terrace, with the help of his friend Dick Italia—and aside from a handful of knick-knacks, my father's work with wood finishes does not alter the functional design of the piece upon which he labors. A chair remains a chair, a table a table, a piano a piano. You might not care to use an item as before simply because, through use, you might mar his restoration, but this is part of his job—to *help* you care. In my work, *this* work, the printed book remains a book, but what I do between and through its pages will seem mightily to alter its form—a place become a sentence, a sentence a document, a document a book, a book a place. And in this living place of present memory a young man in relative poverty becomes an engineer in relative security, an engineer in relative security becomes a writer of uncertain relations. Electronic text itself complicates further this matter of form and differential place, making of "communications medium" both premise and promise—a past and present of here and there, a future of new products, new productions, near and far interrelations of solitude and community. Perhaps new processes of old, new forms of old feelings, old feelings in new trappings.

My father's chief recreation is watching TV, and wood TV cabinets and consoles play an important role both in his employment history and his vocation—all those years spent fixing damaged sets for General Electric, for his friend Johnny Palamino's TV repair business. This is General Electric—not the GE of today, that blue-chip stock that farms out its electronics manufacturing to Thomson Consumer Electronics. But in the end, what's the difference?

You're lucky, Joey, if you can make a living off of what you love to do.

He says this even as the notion of TV-as-furniture has dated badly. It will be years before the emergence of professionally polymeric, surround-sound digital home entertainment centers will recuperate this presumed center of a "living" room. So many sounds and images, courtesy of innumerable sponsors—these mixed signals seem oddly to offer us the remote chance to attempt something more than mere escape: to assign value to our liberties, each and all.

But thinking of him, sitting there on the couch, smoking, watching TV, often cussing at what he sees and hears, *talking back*, and anything but wooden — I swear, he seems the spitting image of himself.
Forget the image.
All works, wood to word, are assigned value, good and bad and in-between, even as all toil under the sun.
One thing is certain: my father knows how to work, how to pace himself. This you learn the hard way, the only way.

Like I say, I love tomatoes. Always have. Except for a two-month stretch one summer, twenty-five years ago, following my junior year at Syracuse University.
We're as broke as ever, I'm looking for a summer job. Mike is in Schenectady, doing hardness testing for the big General Electric plant there. A job my mother manages to arrange for him. So for three months, it's just me and my old man.
I apply for something like thirty-five different jobs. No luck.
I spot in the newspaper an ad for encyclopedia sales people. What catches my eye is a row of $$$$$ across the top of the ad. I figure I may as well see if I have the right stuff for the job. For the $$$$$.
The next day finds me walking out of the middle of a high-intensity sales pitch, pitched to a roomful of poor souls. Walking out into the bright summer air. May as well be selling Fuller brushes, or vacuum cleaners. No fucking way — poor soul or not, it's just not me.
Next thing I know I'm walking into places, cold, asking if they need any help.
One day my father and I drop in on my grandparents, down on State Street. After a plate of chickpeas and rice, I walk out front. I notice for the first time the big red brick, single-story warehouse a block away, S & S Mondo's Produce. I walk in, spot a guy, walk right up to him and ask him if he needs any help.
As a matter of fact, I do. Just lost a guy. When can you start?
Tomorrow.
OK. See you 7:30 then.

Noon a week later, I'm going at it like a house on fire.
I'm standing at the end of a conveyor belt, right at the edge of a seven-foot-high, six-foot-wide machine that's spitting out eight-inch-long

shrink-wrapped packages of tomatoes, five to a package. The machine makes a CHCHUNK noise as it drops each group of five into its green plastic basket, dipping the basket down onto the plastic wrap, cutting the plastic wrap while turning and wrapping the basket, and then placing the package onto a conveyor, where it passes by two blow-dryers, one on each side of the conveyor, which immediately melt the wrapping to seal each end.

    I'm twisting left to grab two packages with my right hand, and twisting right to place them — or more like toss them — horizontally into the bottom half of a cardboard box. Then I'm twisting left to grab three packages with both hands, wedging one in the middle, and twisting right again to toss these three into the remaining space, almost simultaneously grabbing the lid to the box, furthest to my right, with my right hand. Placing the lid on, fast as I can, I grab the full box with both hands and swing all the way around to place the box on a pallet behind me, turning back around to grab a box bottom on my right, and then to my left to place the box bottom in front of me and continuing on to grab two of the three or four packages that have popped out of the machine in the meantime, repeating the process.

    CHCHUNK. CHCHUNK. CHCHUNK. CHCHUNK. CHCHUNK. CHCHUNK.

    I've been doing this now for three hours straight, without as much as a pause. And for a solid hour, an older, well-dressed man, sporting neatly trimmed hair and a neatly trimmed grey moustache, has been standing to my right, eyeing carefully my every move. This is Mr. Mondo, a man my father might describe as dapper.

    On the other end of the conveyor belt, on the other side of the machine from where I work, are four women, each of whom has a basket of tomatoes in front of her, each of whom is sorting through her respective box and placing the better tomatoes on the conveyor belt. Patty, Mary, Josie, and Bernice. These women are all in their mid- or late fifties. And they talk while they sort.

    Mary, wanna go have a few with me at lunch?

    Sure, Patty. Can't wait.

    Yeah, sick of this shit already. Sometimes this machine drives me nuts.

    She's right. Input or output, you're at its mercy. Suddenly the machine growls, spits out a package sideways, crushes three tomatoes, and mashes one plastic basket into the works. Sometimes I can pull things loose, get the machine back up and running.

    I peer my head in. Clearly, this is not one of those times.

    Mr. Mondo walks over and turns off power to the machine, yelling over to a chubby younger guy, his son Sal, to come and un-jam things, and readjust the mechanism. Sal is a decent mechanic when it comes to these machines, but not much else. I've already had to show him how to tune-up his new truck.

So I'm standing there, soaked in sweat, the women have stopped sorting and started talking with one another. Mr. Mondo turns to me for just a moment.

Pretty good work.

He says this softly, so softly it's barely audible.

Thanks.

Let's take lunch.

Lester, the forklift driver loading the pallets into the truck behind me, asks me if I'd like to grab a hamburger with him across the street, in the small tavern. A real dive.

Sure, yeah.

Gimme a minute to finish loading.

Lester is a guy about my size but leaner, in his late thirties, bearded. Doesn't say much, comes off as considerate. And he's what he appears to be — a hard, earnest worker. And nobody to fuck with. He tells me a story one time about how some guys were fucking with him in a tavern, and he walked back out to his pickup, and walked back in the tavern with his 12-gauge pump.

That shut 'em the fuck up.

By the time we walk into the bar across the street, Mary and Patty have already downed a tall bottled Bud each, and are well on their way through their second. They order two at a time each, and place three orders during that lunch hour. They talk about their grandkids.

Lester has warned me to watch myself with Mary, who is less than five feet tall and whose shoulders are three feet across. Her sister and daughter and mother stop by one day. The four women are identical in stature.

Not that you couldn't handle her, Joe. But when she's drinking, you never know. She pinned this younger guy who used to work here up against the wall, wouldn't let him go. I had to break it up.

Yeah, OK. I'll keep it in mind.

When our lunch hour is up, we walk back over to find out how many orders have come in. If a lot of orders have come in, we may be at work till ten that night. If few orders have come in, they'll let us go early. If we've worked a lot of nights, and if the orders aren't too heavy, they'll still let us go early — they don't have to pay overtime, no matter how late we've worked on a given day, unless we break forty hours.

Mr. Mondo informs us that a lot of orders have come in. We're in for the duration.

Around nine o'clock that night, Mr. Mondo tells me that he'd like me to go with Lester to the farmers market and to the P & C warehouse, to help Lester unload the truck.

OK, sure.

Lester makes a quick phone call, and we get down to Park Street at half-past nine. Lester's wife Deb is waiting there for us, a six-pack in her hand.

Let's unload first.

OK.

We unload the boxes, five at a time. Takes us a good half hour.

Here you go, Joe.

The three of us have a beer together. Small talk.

Be home around midnight.

OK, hon.

Lester and I get back in the truck, drive up Park Street to Hiawatha Boulevard, and down Hiawatha to the 690 entrance. We take 690 around the lake to the P & C warehouse. The lake looks pretty at night—smooth and still, the headlights on Onondaga Parkway along the other side of the lake reflecting now and then off the water.

The P & C warehouse is huge. Lester backs the truck up, with me guiding him. We both walk inside, and Lester gives the dock supervisor our delivery order. This portion of the warehouse is refrigerated, so we work as fast as we can.

By the time we get back to State Street, it's pushing eleven o'clock.

See you tomorrow morning, Joe.

Right, see ya Lester.

I hop in my car, a '68 Caprice, and drive home.

This goes on for two months or so. I'm making pretty good money, at $3.50 an hour, plus the overtime I work.

Some days my father drops me off at work, and I phone him when we're through to come pick me up. One evening he arrives just as my boss walks out front. I introduce him to Mr. Mondo.

The two men are about the same age, both first-generation Italian Americans, both smoking cigarettes. My father is respectful, as always.

Hello Mr. Mondo.

The two men shake hands.

I notice that Mr. Mondo is wearing a brand new pair of Nettleton wingtips, a fine pair of wool-blend trousers, and a neatly pressed short-sleeve oxford. I see that my father is dressed much the same way, his wingtips scuffed and worn, his shirt dotted with finishing stains, a few threads unraveling along the cuffs of his trousers.

I see.

I don't know when, exactly, but sometime during this period the smell of raw tomatoes begins ever so slightly to nauseate me. It may have something to do with the garbage cans in the warehouse, which I have to clean out every week. They're encrusted with a sweet-smelling, moldy layer of rotted tomato. Whatever the case, I find that I just can't stand to eat the raw material.

I can recall, when I was a kid, picking tomatoes right off the vine along the side of our house. I'd have a saltshaker, and eat them right there, on the spot. My mother was always smiling.

The tomatoes we sell at Mondo's, most of them, are OK. Some are not OK. Sometimes we can't sell what we purchase.

Tom, I want you to place those pallets in the back room, and gas 'em.

OK Mr. Mondo.

In the back of the warehouse is a small air-conditioned room, fitted with nitrogen. The nitrogen keeps the tomatoes from spoiling, even as they gradually become tasteless.

Tom does what he's told. He speaks with a drawl, is difficult to understand, talks to himself at times, and walks all twisted-up. Lester and the women say that he's "simple," but I'm not so sure. Once in a while someone will crack a joke at his expense, but usually he's left alone. He's a hard worker.

Some days Tom's wife drops in to see how he's doing. I notice that she's not like Tom. Nor are his two kids. But I never make any effort to get to know Tom—I keep to my kind, or to the kind I perceive myself to be.

Our best tomatoes are what we call *chefs*. Each is wrapped in a small, square piece of tracing paper with the S & S Mondo logo, for effect, and then placed by the women individually in boxes.

Be careful with those tomatoes, ladies.

I can hear Patty whispering to Mary, under her breath.

Go on you old decrepit bastard you.

End of July, orders trickle off, and they let me go for good.

It's right after this that my grampa dies, in his sleep. The official cause of death is a stroke, of "natural causes." But he dies at roughly the same moment that Besden's Furniture, less than a mile away in the heart of downtown, explodes with a terrific concussion, so powerful that its huge storefront sign is propelled across Salina Street and into the parking lot of the new Herald-American/Post-Standard building. The explosion is officially determined to be "of suspicious origin." My grampa is 87.

My gramma and my father are visibly upset, and my father consoles his mother. I'm not sure what to say or do. Mike and I attend the funeral parlor visitation hours, meeting a number of relatives we've never before met. My father tells us not to bother coming to the wake, which has been arranged by his three brothers and their cousin, Francesca, a woman who has money, and a strong sense of social propriety. Mike and I are relieved.

When my father gets home from the wake, he's slurring his words slightly, and I notice that slight odor. Been drinking, heavy.

How was the wake, Dad?

He tells us that the long-standing bad blood between him and my uncle Frank has finally surfaced—at the wake.

So everybody is eating dinner, and I tell Sam, quietly—Sam, don't you *ever* volunteer me to drive you and your friends all over Christ's creation. And Frank, he's sitting next to Sam—you know how he is—he pipes up and says, to the entire table, *Why you good-for-nothing asshole you.* So I turn to Frank and tell him, Y'know Frank—one of these days—one of these days, bad arm or no, I'm gonna knock you on your ass but good. *Oh is that a fact?* he says. Yeah, I says, as a matter of fact—

My father jumps up off the couch.

—as a matter of fact, this is as good a time as any, I've taken your SHIT for the last time—

My father doubles his fists now, fuming.

—C'MON YOU SONOFABITCH, I tell him, and he gets up comes at me saying *Why you* but before he says another word I grab him and toss him on the floor I coulda CRIPPLED him, and Sam and Dominick jump on my back and I toss them offa me and to the three of them I says C'MON! And I walked right the FUCK outta there, the hell with Francesca and her BULLSHIT!

My father is furious, shouting, his entire body giving voice to his narration, and Mike and I are trying like hell not to laugh hysterically.

Fifteen years later, when my gramma turns 90, Francesca will throw a birthday party for her, invite a local monsignor. Mike and I will go, and our three uncles will be there. My father won't show. And he won't be around to attend my gramma's 100th birthday party.

Shortly after the wake, the Caprice starts to go bad. I've sunk what money I can into it, but I can see that it's rapidly on its way to becoming history. My father has been out of work for a month, so we've had to let repairs go. Bad brakes. Really bad—takes fifty feet to stop at 20 mph. So it's touch and go driving. And this shitbox puts out more smoke than a chimney. If you get pissed at another driver—as my father often does—you can bury them in a white-blue cloud.

To complicate matters, no money for insurance, or for registration. So four eyes out for the cops at all times.

Even grocery shopping with this shitbox is becoming a source of considerable stress, and my father comes as close as ever to looking into work at Marcellus Casket Company. Marcellus caskets are locally renowned for their beautiful woodwork and finish, and the company pays quite well. But my father just cannot conceive of putting himself to work on that which will be buried in the earth, and he literally wrings his hands over this. The way he sees it, it's working for the dead, not for the living.

When I die, just throw me in a plain pine box, that's all.

He'll get something a bit fancier—but Mike and I will respect the sentiment behind his wishes, and we won't have the money anyway, so we'll opt for metal.

Maybe a TV cabinet and a casket are alike, finally, something my father just can't seem to get a handle on. Inanimate objects, inanimate people. Does it have to be this way?

I land a job pumping gas part-time for a west-side accountant, Dennis Needham, who's trying to make a go of it as a businessman. Dennis is a real prick to work for—rides your ass every chance he gets. He's the kind of boss who'll ask you, casually, to sweep the lot if you have the time, then stop back in two hours to make sure you've swept it. He's always out of sorts, always high on his accounting horse. Even accuses me of not knowing what an "expoTential" is—that's right, this is how Dennis says *exponential*—simply because I have a difficult time using his late-nineteenth-century adding machine.

My father is dropping me off each afternoon at the station and picking me up evenings. Like I say, it's touch and go, and I figure after two weeks that it's not worth the effort, especially not at $2.50 an hour. Not even at $2.60 an hour, which Dennis says he'll pay me if I wear regulation blue polyester trousers instead of jeans. *It's your choice*, he says. But I'm so sick of the bastard that I figure I'll push him on this, and if he gives me any shit, I'll up and quit. We'll see who's the *real* prick.

So one afternoon I push him.

Where are your blue pants?

Sitting over there, on the windowsill.

I point.

But I didn't have time to put them on, Dennis, so I thought I'd just wear my jeans.

Well I don't give a shit *what* you thought—I want to see you with your blue pants on.

But you said this was up to me—*my* choice.

I don't give a shit *what* I said. I want to see you with those pants on.

Uh-huh. Well then, Dennis, that's OK—*I quit*.

Just like that, and the pumps are full of customers. Feels good seeing his jaw drop.

A week later Dennis shuts his place down—he tells me when I phone him that he was losing money—and I stop out to his home in Camillus to pick up my last paycheck. Nice place, with a private office in the basement. He apologizes for his poor behavior, ends up writing me out a check for a few bucks extra.

It's like my father says: you treat assholes like assholes. It's the only way to earn their respect.

My father gets a lead on a job just this side of Solvay, on the west end. A place called Craftsman Interiors.

We drive there together. From the exterior, it's not much of a place. A beat-up storefront that serves as an office of sorts, and a loading dock out back.

My father walks in, and comes out a few minutes later.

We've got the job.

Turns out that I'll be making half my father's wage, which is just shy of ten bucks an hour. We're both under the table. They want us to refinish all the dorm room furniture from Ithaca College. We're to set to work immediately.

We drive around back, park the car, and walk with our spray guns and a few supplies up a few steps and through the door alongside the loading dock into the once-and-future finishing area. It's a large room, perhaps fifty feet square, dust all over everything and dimly lit. Before us are a couple of sawhorses, three fifty-five-gallon drums of clear lacquer, lacquer sealer, and thinner, and a large industrial air compressor and tank outfit in the corner, pneumatic lines leading to the work area. In the background, through another set of doors, we can hear table saws, band saws, rotary sanders—this is where Craftsman actually makes commercial furniture—bars, wood storage cabinets, and the like.

I can tell that my father—who is not a patient man—is not pleased with this arrangement. The workplace is a mess, and it's abundantly clear that, prior to our arrival, there has been at best only a half-hearted attempt in these confines at laying anything resembling a finish on something resembling wood. It's a wonder this guy, whom my father refers to simply as James, was even awarded the Ithaca College contract. My father begins to arrange the sawhorses, but gets so flustered after a few minutes that he kicks one over.

C'mon Joe! Let's get the fuck outta here! I'm gonna tell that no good Arab bastard that he can stick this job up his ASS!

Dad, wait a minute—we've come this far—

Gradually he calms down, and within the hour we're moving the small dressers and drawers from an adjoining room into the finishing area, lining them up on the floor and preparing them for a finish.

The furniture that we're working on has been stripped in another area. Our job will consist of sanding the surfaces, prepping them for sealer, and after an undetermined amount of sealer, laying on a coat of semi-gloss lacquer. We'll be working on something like eighty drawers, and forty small dressers, at a time. It's up to my father to decide what constitutes adequate sanding, sealer, lacquer—adequate for this sort of assembly line, production work. An assembly line that runs by hand.

And it's also up to James, because James leans on us—hard sometimes. He's got a deadline—end of summer. And he wants to beat his deadline. So he spies on us, peering through the crack in the doors that separate our area from the furniture construction area. He's constantly coming into our work area, checking up on us. James Mahmud doesn't seem like such a bad guy, actually—a handsome guy of around forty, married, with kids, who also owns a restaurant at Armory Square. But he behaves like most entrepreneurs I've run across—as if you're stealing their money, jeopardizing their undertaking. So it's only a matter of a few days before it becomes my father the Sicilian against James the no good Arab bastard.

At first I learn by mimicking my father. Sometimes he tells me things I already know, things I've heard all my life. But I try not to object—I try to learn.

Hold the block like this, Joe—and move it with the grain.

Joey—move the gun back and forth steadily—not too slow, the paint'll run!—release the trigger quickly and press again firmly as you reach the end of each coat. That's it—

I learn, as I've learned on other jobs, that I need to relax into these muscle movements—I need to let them become a part of me. The rhythm is a pulsed whisper, a shhhhh/shhhhh/shhhhh/shhhhh, and a swinging arm and hand becomes visual accompaniment. As it is, my hands and forearms tire quickly. I'm tensed up, my reflexes haven't yet established the right feel for the work, for the material.

But after a week, my father tells me that I'm as good as any pro. It's not exactly an overstatement—I have a great teacher, and I learn what corners can be cut without sacrificing quality. The gun becomes an extension of my hand, and I adjust the spray width to suit my own peculiar motion. I even monkey with the compressed air–paint mixture to determine what seems the best setting given the mixture of paint and thinner we've made for a given batch, and the humidity and temperature on a given day. Eventually I begin to decide for myself when a piece is finished. I begin to make this work mine, and it happens quickly simply because we're doing so much damned work, and because my father trusts me.

One thing I don't need my father to teach me, perhaps because he's already taught me, by example: when to sound off, and at what volume. And once in a while, I do—but good. I finally tell James just what I think of him spying on us.

What you mean?

I saw you over there—spying. Who do you think you are?

Who you think *you* are?—I own this place, I'm the boss.

There's something almost endearing about this guy. He's like a spoiled kid, uses baby logic on us. But he's got *money*—

Joey! Listen, James—I don't care if you're the boss or not. You go away and let us do our job.

James storms off.

Bastard!

In addition to my father and me, James has working for him a handful of cabinetmakers: a Greek guy, Pavlos; a Sardinian, Armand; a Jewish kid, Jacob; and a few other guys who keep to themselves, and who speak hardly a word of English. During breaks Pavlos will walk over, and we'll shoot the shit together. He talks under his breath, and quickly, is a bit difficult to understand. But after a while you pick up his accent—or more properly, his inflected English.

He's a talented woodworker, but wants to open his own bakeshop. So he's working his ass off, saving his money. Like a lot of immigrants I've known.

Once in a while we'll all get together after work, before heading home. The shoptalk is animated and confusing, the place a circus of hand signals. And somehow out of this tangle of dialect and gesture, work gets done. Good work.

As always, I make mistakes. And my father covers for me.

After finishing a large batch of furniture, we'll drive the company truck—an eighteen-footer—around back and up to the loading dock, and carefully stack the furniture in it. Takes an hour or so. Then we drive a quick mile over to the warehouse, where we unload the furniture with the help of a few other guys. James will store the furniture here until the job is completely done.

Joey, why don't you drive today.

OK Dad.

Today I decide to have some fun, make it a *really* quick mile. I gun the truck this way and that around corners, my father yelling at me to slow down, take it easy. But I'm having a good time of it, and I can see that he's smiling too.

When we get over to the warehouse, James and Pavlos and a few others are standing outside, waiting for us. It's a beautiful, warm day. I back the truck up, and get out and walk around back to open the rear door. As I release the door and it flies upward, I jump back. The entire load has shifted backward, and is hovering over the edge of the truck, about to topple onto the asphalt.

Nobody moves. Nobody knows what to do. Suddenly a drawer drops off the top of the tottering wall. I reach out and grab it, catching it in my right hand. Then another, which I catch in my left hand. And then a wall of drawers. I jump out of the way, and the entire heap comes crashing to the ground, smashing the drawers, some of them reduced to splinters.

James is furious. Pavlos and the other guys are silent.

Me, I don't know what to say. As it is, I have a hard time not laughing. And if *I* say I'm sorry to James—whatever he thinks of my apology—I'll implicate my father in my own foolishness. Because it was my father's decision that we would take turns driving.

But my father—he's serious, resolute.

I'll take care of it, James.

Nobody says a word. Not a word. We unload the rest of the furniture. My father and I get back in the truck to drive back. He's driving.

Sorry Dad.

Just take it easy next time, Joey.

Back at the shop, my father sets to work after hours, with his clamps and glue. Within a day, he's reassembled the broken furniture, touched it up, refinished it. Good as new. In fact better, sturdier.

One thing: while my father and I are spray painting, you can't see ten feet, the spray fumes are that thick. And there's no exhaust fan. All we can do is open the overhead door, and hope the fumes drift outside. James never visits us during these phases of the operation. Nor does anybody else.

In fact my father has *never* worn a respirator—spray mask—to help filter out the fumes. In fact he smokes *while* he works. In fact I believe that smoking has increased his *apparent* tolerance to fumes. In fact it *is* a bit difficult to work with a mask wrapped around your face. But in fact you *can* get used to it. And in fact you *should*.

Still, fact or no, should or no, asthmatic or no, I decide to follow in my father's footsteps, knowing little of the science of solvents—which solvents are soluble in body fluids, which are not, and what their effect on the body may be.

* * *

I should have taken an educated guess.

Reminds me of when, as kids and like so many of our generation, we'd chase behind the mosquito spray truck as it made its seasonal runs down Dolores Terrace—a couple of evenings in early summer right after prime time. This was before they installed streetlights. We'd play hide & seek in the aromatic plumes. I'm still awaiting the results.

Doesn't take so long at Craftsman Interiors. Four weeks in, around the time I spot and swipe an old Binks 7 from the shop—bring it home, clean it up, rebuild it, fuck you James you were using it for *glue* for Christsakes—I begin to develop a slight pain while breathing, on my right side. At first it goes unnoticed—I'm not sure what I'm feeling, think it to be a muscular pain. But shortly after the pain surfaces, I detect as well a faintly bad taste in my mouth, especially when I cough from the fumes.

And one day, one cloudy rainy day, I begin to get the chills. I'm standing near the overhead door, and the cool damp air is making me shiver. And I have a knit cap on my head as we work.

The next day I'm in a sweat, and it hurts like hell to breathe. My father doesn't mess around, takes me right to the doctor. Who himself doesn't mess around, has an X-ray taken.

It's pneumonia—a classic case, as I learn from the books I'll read years later. Books that warn of the hazards of working with such solvents. My father—after work each day, he spends five minutes blowing the black shit out of his nose onto his ever-present hanky. And he sneezes six or seven times hard every couple of days, kicking the shit up and out of his system. Doesn't seem to bother him—not yet anyway.

I spend the last two weeks of the summer in bed, weak, feverish. It's a slow recovery, even with my father's full repertoire of dishes—chicken soup, macaroni, escarole, chops—even with the penicillin, a substance that I'll come to understand better when my studies lead me out into the working world of engineering professionals, anaerobic and aerobic fermentation processes.

Sometime during my recovery, I regain my taste for raw tomatoes.

It's mid-decade, and the public era that will in retrospect be stereotyped as *the seventies* is in fact just beginning. A new president sits in the White House, and the political upheavals that served as the backdrop of my first two years of college are all but over.

No more public affairs courses with Professor Cope, the registrar of the university, a wise old man who understands the law, political action, and social reform, and who holds classes in his conference room. Cope strikes me as a conservative politically, but he always challenges me to think through, and to commit.

No more English—even if all I can recall is a dry discussion of Whitman's poetry, and a few lines from "Howl"; no more history of religion, which opens my eyes to the multiple authoring of the Torah; and no more philosophy courses, which I love for what I find at the time to be a certain purity of thought. I no longer have to take any but math and engineering courses, which are often more than I can handle.

There's talk of Hank Aaron retiring. Aaron, my last sports hero, a guy who can hit the long ball off a bad pitch—they say he has strong wrists. It's something I aspire to, but my wrists are on the small side.

Saturday Night Live is making a splash, the word *video* is on the tips of more and more tongues.

And school is eating up more and more of my time, the disjunction between my life of the mind and my life at home growing more pronounced with each new semester. That the old wooden structures lining College Ave. are being torn down, that much-needed, but more monolithic edifices are replacing the older, itemized accretions of the campus—no matter. I can see the university only in terms of its relative charm, its embodiment of what seems to me, green as I am to a world cultivated so, a finer set of symptoms. I am grateful to the university for this golden opportunity, and yet—

It will take another two decades of gutting out faulty social and secular and free-market institutions before I'll realize, still hopeful—as hope springs eternal, and dies last—that charm is exactly what the trustees of the university *want* me to see.

I'm beginning to hang out with a few of my classmates from my engineering courses—Gerald, Kent, Bill. And once in a while I'll have a beer with a guy I meet in the campus gym, where I work out with weights—Steve, an electrical engineer, or what we call a double-E. Steve and Bill are townies, like me. Gerald is studious, but a bit of a tightass. He rooms with Kent, who has a good sense of humor, and hails from Poughkeepsie. Kent's folks are—according to Gerald, who says it each time with resentment— filthy rich.

Mike is himself a sophomore in the engineering program at SU, and he and I will head up to campus on the odd Saturday to party with some of our friends, or to crash a party. Because of my uncertainties about settling down, my relationship with Julie will falter—again—and I'll begin to go sweet on a poli-sci major I meet, a Jewish girl named Hilary, who plays guitar and sings "Anticipation" with feeling. Not much will come of this flirtation, though, and Julie and I will get back together. As commuters, Mike and I will remain somewhat removed from campus activities, campus life.

Back on the quad and still a bit weak from the pneumonia, I stand among the throngs who watch the streakers streak by. Six men and one woman, all white, extremities bouncing in the cool autumn air. Activist energies rechanneled? I wonder what my father would make of all this.

Winter drags, courses drag. In the spring, I receive in the mail a small manila envelope. Inside, a fifty-page pamphlet, off-yellow cover, accompanied by a letter.

Dear Fellow Engineer:

As a senior in Mechanical Engineering you should
soon be entitled to the privilege of
promotion to the Associate Member grade of The
American Society of Mechanical Engineers.

The Old Guard is a group of dues exempt
members of ASME dedicated to assisting the student
to become an active Professional Engineer.

As an initial effort in this direction, The
Old Guard offers you a copy of "The Unwritten
Laws of Engineering" by W. J. King, and we
recommend a careful reading of this pamphlet as
it will help you understand that professional
relationship which makes a man a real engineer
in his community.

Do not be put off by the language of this
paper. It was not written today. But like the
Declaration of Independence or some of Lincoln's
speeches, the intent or philosophy of the
writing applies equally well today as it did the
day it was written.

We trust you will enjoy it and profit
from an application of its teachings.

                                          Sincerely,

                                          Chairman
                                          Old Guard Committee

* * *

I flip open the pamphlet. "*Copyrighted* 1944 *by* The American Society of Mechanical Engineers *when published in May, June, and July,* 1944, *issues of* MECHANICAL ENGINEERING." I browse through, spotting a few lines toward the end that give me pause:

> It is very much like the design of a piece of apparatus. Any experienced engineer knows that it is always possible to secure substantial improvements by a redesign. When you get into it you will find that there are few subjects more absorbing or more profitable than the design and development of a good engineer! As Alexander Pope wrote many years ago:
> "The proper study of mankind is man."

The proper study of mankind: I think of D-day, months prior to my mother and father meeting, in Europe. I think of history, private and public, of my father's palpably creative process, *then*, and of my intangibly mimetic representations, *now*. And I think of the work of art as an artifact in the offing. Whatever its appearance, I think it is never finished, I think it does not simply or proportionally represent, or work against representing. Instead, it *presents* in some present moment a past endeavor, making of the past what Gertrude Stein articulated, years before my parents' budding romance, as a *continuous present*. A continual presentation, a presentation that foregrounds—*here, now, mine*, and I trust in some sense yours, and for the future—processes of representation, presence of absence of presence, simultaneity of here and now and then again, loss and gain but above all, that time-worn cliché of enduring hope. The sun and moisture spotting, the yellowed aging, the sheer damage that touches wood and paper and flesh alike, and the urge to renew. Alike in reproducing materially this process of representation—whether as pixel or laser draft or print run or digital distribution or daughter or son or garden growth—a falling ever, ever into the representations and presentations of process, whorls of fine-tuned processing. Endless spiraling of the generations regenerated, steadfast inklings and shimmering non-inklings, surfacing and deepening alike differences of kind and kinship and degree.

And at this moment and forever, for and against me: the tactile more the province of my father, whose hands, calloused yet soft, would that he could rest occasionally on my shoulder. And—it must be added *here, forever*, in species blood-script: the fifties crimp & solder piecework, the seventies and eighties typing, the endless sewing and knitting—of my mother, whose fists would curl inward at death.

9.
Primitive Roots

By magic numbers and persuasive sound.
—William Congreve, *The Mourning Bride*

Definition 2.8 *If a belongs to the exponent ø(m) modulo m, then a is called a primitive root modulo m.*

PRIMITIVE ROOT?

I'm sitting at the kitchen table, book open, staring in.

The lamp is on. The clock is ticking. The refrigerator is humming. The fan atop the refrigerator is blowing first one way, then the other. The oven is on, the oven door is open, the stove burners lit. The TV is on, in the other room.

Joe, will you look at this asshole for Christsakes!

He's swearing at the TV. I can hear a talking head confirming something or other about an energy crisis. My nose edges a centimeter closer to the book.

My fingers are getting cold. I stand up and walk over to the oven, turning my back to the open oven door. The heat feels good. After a minute or so, I sit back down. Mike walks in.

Cold out there.

Yeah.

He takes off his coat, throws it over the back of one of the old wooden chairs at our kitchen table. Same chairs my gramma and grampa once owned. My father walks into the kitchen.

Boys, how 'bout I fry you both up a coupla chops?

He claps his hands and rubs them together as he says this.

No, that's OK Dad. I'm not hungry.
You sure?
Yeah.
Joe?
I'm working.
You hungry?
No.
What's that you're working on?
Nothing.
OK. Remind me tomorrow I gotta pay Freddie.
OK.

Theorem 2.25 *If p is a prime, then there exist $\phi(p-1)$ primitive roots modulo p. The only integers having primitive roots are p^e, $2p^e$, 1, 2, and 4, with p an odd prime.*

Let's back up some.

We're six months behind in gas and electric. They've shut off our electricity, killing the thermostat that controls our forced-air gas furnace, and killing the fan that forces the air. No heat. But we have stove gas, and hot water. Why they haven't shut off the gas we're not sure, and we're not about to ask. We've run an extension cord out the kitchen door downstairs, over the inside stairwell. Plugged into the outlet just below the bulb that hangs above the entrance to Freddie's flat. My father gives Freddie twenty bucks a month.

It's my final year of college. I'm deeply into both my mathematics and engineering degrees, my college life a litany of function and formula.

The math I study for my math major, over and above the math courses required for engineering, is more abstract, less applied. It's the internal logic that's at stake in these courses—definitions, proofs, theorems, and so forth—and judging by my grades in calc, I've developed a better grasp of mathematics as a discipline than have my engineering cohorts. I have a better sense, not so much of how, but of *why* a given set of equations might be appropriately applied to a given set of problems. I'm learning, slowly, to understand various assumptions, postulates, as such, and as an integral part of the curricular whole. And I think of myself, always have, as a slow learner.

I teach myself by reading a theorem and its proof over and over, and trying to describe it, in my own words, to a hypothetical teacher-interlocutor. When I can manage to describe the theorem to the interlocutor's satisfaction—when no further dialogue is necessary between myself and my hypothetical teacher—I've learned it. Perhaps the interlocutor is in actuality the student, *mutatis mutandis*. I can't say.

But my abilities, as I'm finding out to my dismay, are anything but boundless. This is an area of study in which one bumps up against one's limits rather quickly. There's a young woman in one of my classes who seems to see *right through* the theorems, for whom the answers seem to be a matter of mere focus. Math major or no, math scholarship or no, in my prime or no, my math courses are giving me conniptions. And my toughest course, far and away, is number theory.

Number theory: one has first to think in terms of ones, integers, wholes. Subtraction, division, difference. Prime factors. For two integers, a and b, to be relatively prime means that they share only that primary integer, 1—like o, not a prime number—as a divisor, or factor. You write it thus: $(a, b) = 1$.

Does a human integer—the human factor—remain relatively prime to itself, indivisible by all others, regardless of kinship?

Assuming that there are an infinite number of numbers, there must be an infinite number of primes. The proof, Euclid's, is simple, elegant, proceeds by contradiction. Assume that there are in fact a *finite* number of primes. Imagine the number, N, formed by the product of these primes. Now add 1 to N. The result, N + 1, is divisible neither by N nor by its prime factors, hence must *itself* be another prime. Which contradicts the premise, *QED*.

Infinities, of course, may be made to yield all sorts of paradoxical results. Smaller and larger infinities, for example, as Cantor's mathematics of transfinite numbers suggests. We can posit an infinite number of primes, or an infinite number of relatively prime pairs—an infinite number of such differences, so defined. Hence—stretching my imprecise metaphor—for any finite number of beings, there is the possibility that all such beings are relatively prime.

I bet you think all of this is about being Italian, or Italian American, right? So many movies, so many books, everyone knows what this means, or thinks they do. You too, Joe, but I'm not so sure, and it runs in your blood. Or so I'm told. It runs in your blood, they say, along with the French, the German. And as you'll find out decades later, the Jewish.

My interlocutor is extremely unhappy with me at the moment.

Just what the fuck is a primitive root?

My fingers are getting colder. Oven or no, the kitchen temperature is dropping, ever so slightly, by the hour. Pinkie temperature and comprehension are, as they say in the trade, directly proportional. And, I tell myself, the curve descends more sharply as it plummets toward the X-Y(-Z) origin. Another fucking ice age.

Joey, look at this asshole, will ya?
Dad, I'm trying to study.
OK, but will ya look at this asshole.
I get up and walk into the living room.
Turn the channel why don't you?
To what? Three goddamn channels. Commercials every goddamn minute.
Four channels, including public broadcasting.
Ah, garbage.
They show good films every now and then.
Garbage. They say we'll have cable TV someday soon. Thirty-forty channels.
Yeah, some without commercials.
Yeah, what do you think about that?
More choices, technically speaking. But it'll cost.
How much?
I don't know.
Well maybe we'll get it when it's available.
Uh-huh. Listen, I have to go back to studying.
Go ahead Joey. What's that you're studying again?
Number theory.
Je-sus Ch-rist. Tough huh?
Yeah. Real tough.
You'll figure it out.
Yeah.
OK. What's Mike doing?
He's in the bedroom studying.
Oh.
Mike walks into the living room, sits down.
Hey Mike.
Hi Dad.
My father places his hand on my brother's knee, squeezing.
Quit it!
They both laugh.
I walk back into the kitchen, sit down again. But it isn't long before I get up from the table, decide to see if *Britannica* can offer some aid here.

Primitive Roots

* * *

With the money I earned from the prior summer's jobs, I've picked up a brand-spanking-new *Encyclopædia Britannica* set from my father's friend, Cliff Calhoun. Cliff is divorced—a hard drinker, a West Virginia hillbilly, and a really nice guy whose face suggests perpetual torment. He's also a damned good upholsterer, and my father and he sometimes work together, one handling the wood repair, the other the fabric.

But the thing with Cliff—he loses his head every now and again. Gets pissed off at Radio Shack over some equipment they won't warranty. And the wound festers until, late one night, after half a dozen too many, he drives his pickup over to the mall, backs it up to the mall doors, hooks up some chains, tears the doors open, and with security alarms screeching, hustles over to Radio Shack, tools in hand, where he busts open the inside set of doors and grabs two armloads of new hi-fi equipment, hustling back and throwing it into the bed of his pickup, speeding away before the cops arrive.

For twenty-five bucks, I buy the cassette deck he's swiped. Not a bad item.

But Cliff has decided to move back to West Virginia, with his ex-wife and young daughter. And so he decides to sell everything—including a brand-spanking-new *Britannica* set he's received on approval, for his daughter's education, but hasn't put a dime down on.

He's asking a hundred bucks, even. The set is still in the boxes, unpacked. I figure this is a fair deal.

Let me tell you something, Joe: you were raised in what some would call a largely working-class, ethnic city, with specific ethnic communities—and you don't even know it. You don't understand that your life at 112 South Dolores Terrace is in large part a function of a modest income. You rarely hear the word "pasta" as a kid. But you eat plenty of macaroni, and plenty of onion soup. And you love a good sauerkraut. You hear four languages, in fact, only one of which you really understand. And your parents believe that this is how it should be—that this is how Americans raise their kids, that Americans buy white enriched bread to accompany their meals, that this is the stuff of full-blooded American boys. You'll learn as a toddler to mumble certain words most of your friends don't use, like "Oma." Comes easy.

And not until you find yourself in "less" ethnic, better-heeled regions will you begin to notice these things, the college classroom your first real contact with privileged white folks. It takes you some time to adjust, to find ways to "fit in"—and you won't even know you're doing it. You don't "look Italian," and what others will identify as the ethnic in you is comprised of a collection of pronouns—us, them, you, me—twisted around a few verbs, adverbs, a few food groups, a few gestures, maybe even a few superstitions. As you age, you'll tend to recognize yourself as a function of such discriminations, not least because you'll notice the looks you get when you pack a lunch of figs and olives and cheese and Italian bread. It's in your blood, the food, literally. And the words? How does one live up to such things?

If two integers, a and b, have the same remainder on division by a third integer, m, they are said to be congruent modulo m. You write it thus: $a \equiv b$ mod m. This is equivalent to saying that m is a divisor, or factor, of $(a-b)$, or $m|(a-b)$.

Deceptively simple formulations such as these often take some time to adjust to. If I tell you that any integer that is not a prime must have a prime factor less than or equal to its square root, the implications, which are vast, may not be immediately apparent. And regardless of how elemental the operation, most minds resist the appropriation of ordinary language by mathematical symbol. Here, at least initially, one must learn to do away with resistance.

The set of integers congruent to a given integer modulo m is called a residue class.

On a human scale, my father, brother, and I—a, b, and c—may be thought equals. But are we equal? Or are we merely congruent? Is our similarity apparent only when you factor in another, m, from the outside? Could it be said that we are, each of us, congruent modulo m—does factoring each of us by this other, m, yield the same remainder, a similar difference? But *which* other? And are there yet other others—d, e, f, ...—who will satisfy this relation, comprising a *set* of others congruent to a, b, or c modulo m? Is this what is meant by a residue class modulo m?

Clearly, there are m such residue classes.

Clearly. Of course.

This sort of language is a commonplace in mathematics. I'm beginning to wonder about the history of the discipline, about how mathematics developed over time, across cultures. Who started using this sort of language? Why?

The proof is trivial, and is left to the reader.

Why the fuck am I being taught this way?
Should I be asking such questions?
I learn quickly that most of my instructors won't offer much in the way of instruction. Most profs lecture—talk at us, leave it to us to figure out what's

going on, sometimes with the aid of a recitation instructor. In one large lecture section, the prof actually distributes actual copies of portions of the actual textbook, then uses an overhead to mark up actual transparencies of this very actual material. Dismal. In another course, the instructor uses his own textbook, constantly references his explanations there, and refuses to give us test questions of the sort we've been assigned for homework. Excruciating. And in yet another course, multiple-choice questions are used to test your ability to answer multiple-choice questions:

Which of the following statements (A) through (E) are not incorrect?
(D) A and C.
(E) None of the above.

Etc. Most of my classmates congregate in their frats, or dorms, or apartments, and collaborate on their homework. Which is as it should be, finally. Me, I spend a lot of time in my own head, with my interlocutor.

What with the frictions between the college campus and this city, between those who think they want to know and those who think they know what they want, maybe you find a way to translate, Joe. Drop an unnecessary pronoun, turn things around. Repeat yourself. Maybe he does, and maybe she does too. Maybe all of this rubs off on you.

That's why you used to say windowshield, why you still say pork hockies. That's why the fish we're eating has fins, Joe, not bones. That's why you pronounce certain words a syllable at a time. Vay-por-eyez-ay-tion. Why you say oodles, like your mother, and curse with such passion, like your father.

Thing is, with a few exceptions, my best instructors are often the recitation instructors, themselves graduate students. DK, from NYC, in physics. FL, from Egypt, in thermodynamics. My advanced calc instructor AB, an Armenian, is a teaching assistant, and like DK and FL, he knows how to talk it out with you. Some of the students object to the inflected English of international instructors. And some of the instructors do indeed have serious problems with English. But the majority speak English as well as native speakers. And me, I grew up around Italian-French-German inflections, what we call *broken* English—to me, this is the way of all worlds.

Focus, Joe, focus. On the page, on what it's telling you, between the curious ink markings, the signs. Read between the words, through the symbols, inside the margins.

What is it saying to you, Joe, what are you saying to me?

* * *

My brother and father start laughing in the other room.
Hey Joe, Mike asked me where you'd gone yesterday afternoon, and you know what I told him?
What?
He shit and the hogs ate him.
OK Dad.
They both laugh.
OK you say? He asked me where you went and I told him, He shit and the hogs ate him.
My father and brother are laughing harder now.
Joe, explain to your brother what that means.
Dad, I'm trying to study.
He shit and the hogs ate him!
They're positively rolling over with laughter now.
Why are you two JERKS laughing? Can't you see I'm trying to study?
Peels of laughter.
Why don't you two hit the shithouse!
In fact I know why the two jerks are laughing. I'm smiling myself, nose right up against the page.

As a student at SU, I've been allotted time on the mainframe—a time-sharing account. I tinker a bit in Basic, while some of my engineering cohorts boast a great facility with Fortran (aka Fortrash to contemporary hackers). I'm unimpressed—I see little point in mastering the nuances of programming, given that I'm already immersed, if not drowning, in the language of mathematics, exhibit an aptitude for formal logic, and desire if anything to learn my mother's tongue, French. So although I can intuit the immense calculational advantage, even beauty, of programming languages, I rationalize away their utility.
The machines should be capable of handling English. Like HAL.
As I reason it, as a math major, I've honed my ability for handling the logic of propositions and predicates, the basis for intentional, and in this latter sense *artificial*, programming languages—should the need arise. It hasn't. What I don't quite grasp is that exposure to these more specialized languages will no doubt ease anxieties owing to the coming personal computer "revolution." Revolutions are themselves symptomatic of the times, and this one will face little resistance from the establishment.
True, I can't spend any more time up on the hill than I am currently, which is what the mainframe environment will require. By the same token, in order to handle problems in the new SI *and* English System units, both of which are

being used in my classes concurrently, I'm stuck using conversion tables in addition to trig tables and log charts and, on rare occasions, a slide rule.

But the computational zeal of my engineering professors eventually wins out. These guys—and they are *all* guys—have begun to demand ±1% accuracy on homework and exams, significant digits aside. When you couple this insistence on accuracy with complicated calculations that entail extremely small or extremely large physical-numerical inputs—.065 newtons/m3 or 33,800,000,000 ft-lbs/minute—the lesson is clear, regardless of one's feel for decimal points: bringing anything but a calculator to class is fast becoming an uncalculated risk.

So in my fourth year of college, prices drop and my mother can herself afford to buy both Mike and me a Texas Instruments SR50. Now we're as equipped, and as compatible, as our peers. And this widespread use of calculators will, in turn, help prep the consumer market for personal— home—computers, the latter only an idea at this time, but an idea on the cusp of becoming a reality.

Unwittingly, then, these profs are contributing to the coming digital revolution—of the *alphabet*. Their passion for numerical accuracy and crunching numbers helps drive the development of what in a decade will be sold to consumers primarily as *text*-production devices—properly speaking, machine language and assembly language and the like all in the service of natural language-based documents. And all in the name of growing service and office economies.

In his book *To Engineer Is Human*, Henry Petroski illustrates how engineers, in the wake of the pocket calculator, may fail to grasp the extent to which their digital calculations are but approximations of the real. Petroski discusses how corresponding optimizing trends have resulted in structural designs, for instance, that may use materials more efficiently, but that lack the overdesigned structural integrity one finds in the bridges and buildings of bygone eras.

Petroski knows his business, and there is nothing rearguard in his appraisal of older versus newer computing tools; he simply wants technical thinkers to continue to learn from their failures and successes, and this extends to their computational habits.

Still, we sent a man to the moon thanks to computers, not slide rules. Or such is the mid-seventies mindset of the would-be engineer.

And me—around the time ABC is becoming the prime network contender—I've got one foot in Link Hall, with my engineering buddies, only a woman or three in the lot. And the other foot in the old Carnegie Library, the building that houses the math department and the Engineering and Life Sciences Library. And with this other foot, and in addition to real and complex analysis, I've stepped square into number theory—the language of numbers.

Years later, I'll return to the engineering library to read all about the history of my chosen profession. Not math—when I graduate, the only company actively seeking math majors will be the CIA. Not my line of work, not a chance. And in retrospect, my "chosen profession," engineering, will reveal itself in so many ways as chosen *for* me—by the marketplace, itself zoned by powers beyond my ken, powers of the third person, that simply and utterly *be*.

Spoken like a true first person. My eye, you say. So what *about* you? Where are *your* loyalties, who taught *you* how to count 123 to 10, and back again?

I feel a residue here, an "us" vs. a contingent—

Powerful imperatives must be addressed, yes? Me, I was raised like this, or something like this. Could handle simple arithmetic before school, schools.

Still no good in water though, at once in and out of my element, elemental.

More years will pass before I explore the alphabet using those selfsame computer technologies, generations advanced, that I'd avoided as an engineer. As a poet, I'll immerse myself in the arcana first of the Net, then the Web, dabble in HTML, try to tease out of the telephone lines and eventually the sky itself those distances that isolate, separate, and bind together, all via a sequential alchemy of 0s and 1s, those non-prime digits that would presume to render the realities of concrete or electron as easily as the mysteries of flesh.

> *Fuck 'em if they can't take a joke.*
> *Only once you run the pork and beef through the grinder, only once.*
> *They all carry knives in their back pockets, Joey, you gotta watch those little guinea bastards right off the boat, you gotta watch 'em.*
> *Grampa and Gramma were ignorant people, boys. Not stupid—ignorant, they were ignorant, simple people.*
> *Believe you me the Germans will start another war.*
> *Jojo, go get Mikey, go get your brother.*
> *Left a lot of meat on the bone, he says to Mike, that's the best part.*
> *Goin' to town on that steak, huh Joey?*
> *All that woman would feed us, she says, is pasta fajioli—pasta fajioli, pasta fajioli every day when I came to this country—after the war, mind you—and what did I do to deserve that?*

The proof is in the pudding.
Joe, he says to you, you talk like a man with a paper asshole.

This is African American dialect originally—paper ass. And this first-generation, white, Sicilian, North American, divorced, arthritic, alcoholic WWII veteran is telling your white ass something: you talk a good trade, Joe. You talk like a man with a paper asshole. Your mouth moves like a whippoorwill's ass. And your ass sucks canal water.

Language drifts, paper ass to paper asshole, asswipe to canal water. The stuff of life, the stuff of neighborhoods, of appetites—under the weather, across the vowels, against the sky.

It's your life, and you must comply. Amateur or professional, there's a rock with your name on it at the end of the road. You'll need all of your arts and all of your technologies just to keep your head above water, just to keep the literal from foreclosing on the real. And what then?

Like your father always says, Joe, shit and two makes eight. In the world in which you're finding your persistent self, everything adds up. Seven and three makes ten.

In your brother's December 1976 issue of High Times: *an article on the Burroughs 7700 computer, "the heart of TECS—The Treasury Enforcement Communications System, the information-retrieval network behind the soaring dope seizure and arrest statistics"* (39). On the home front: your brother can teach you a thing or two about brewing beer, or growing pot. On the horizon of your professional expectations: a seven-year stint in the brewing and drug industries, a developing expertise in process control.

Laws of nature, or of culture? Ecologies, or economies of scale? The world as you find it? Or the world as it finds you? Little wonder some seek out alternatives, and some rail against the status quo.

When things add up, it adds up. And it smarts.

The proof is trivial, and is left to the reader.

In an instant I've got it. My internal dialogue has configured the logic of numbers into a structure of mutually dependent assertions. And I can grasp at once the conceptual pattern that has emerged from these varied assertions—the whole nine yards, as they say. Three and a half hours after I first crack open the book, my interlocutor signals satisfaction, *QED*.

The next morning the floor is so cold when I tumble out of bed that my toes go numb on contact. I hop into the shower for a quick rinse, throw some clothes on, Mike does the same. The three of us get into our current shitbox—another '69 Camaro, this one originally Mike's. Three-speed manual, steering-column shift. Clutch has just about had it, pressure-plate is so-so. In another month or two, we'll have to floor it, get a running start, and say our prayers to get up a three-percent incline.

The streets have been salted and sanded heavily, so my father makes it—just—up to the top of the hill, where he drops us off on his way to work.

Have a good day, boys. See you around six, OK Joey?

OK Dad. See ya.

See ya.

Mike splits off on his way to chemistry, where he's having problems of his own. I get to class a bit early, and take my seat toward the back of the room, huddled into myself to keep warm.

I open up my textbook to refresh my memory, quickly satisfied that my insight from the night prior has carried over, intact. The other eight students trickle in. The last student to enter is my pal from Palestine, a grad student in math named Ahmad—unshaven, wearing fatigues, intensely serious, always greets you with something of a scowl.

After class one evening we exchange a few words about how difficult the course has been. It doesn't take us long to determine that, when it comes to U.S. English four-letter words, we share pretty much the same vocabulary. Ahmad even teaches me a few Palestinian equivalents.

We give each other the high sign, and he sits in the empty chair to my right.

Hey!

What?

We're whispering loudly. A few of the other students turn and look at us. Ahmad glares back at them, glancing at the classroom door nervously. They turn away. Ahmad is leery of the prof—a staid, white, middle-aged man whom he neither cottons to nor trusts. He's worried that the prof might catch him asking an undergrad for help.

He leans over to me, eyes darting back and forth, wild black hair touching my shoulder.

Hey—what the FUCK is a primitive root?

10.
Say-Cursed Susan B. Anthonies

> Neither blessed like the elected, nor hopeless like the damned,
> they are infused with a joy with no outlet.
> —Giorgio Agamben, *The Coming Community*
> (trans. Michael Hardt)

AGE TEMPERS YOU. Hard as need, soft as help.
She gave him two sons, he gave her two boys.
They'll never be grandparents—not in their lifetimes. Maybe not ever.

Back from the instant of death, from organism's zero energy, taken whole. How to reckon, how not to.
My father, Joe—Giuseppe sometimes to his folks—and my mother—Suzette (Sue or Suzie to my father and his family)—were both born in 1922. A Capricorn, a Leo, then a Gemini, followed by a Libra—my family, ordered by birth.
I wait while she dies. I can see her body, sitting up, straining.
She's still wearing her small, gold earrings. Mike removes them, gently.

Drawing a paycheck, salaried, life-preserved: not even off-white, not one hundred percent anything, and not fair? What's in a vowel, finally, or difficulty in pronouncing a name? Moving the cursor, stirring the sauce, working the clock, alarm set to go off—
I watch him die, his eyes rolling back into his head. I'm not there, but I can see his final look, the one he gives us as kids, the one he gives us for a lifetime.
Don't shit yourself.

* * *

Susan Brownell Anthony died in Rochester, NY, in 1906. She was 86.

From temperance to the voting booth to the grave—a conservative abolition.

12 chapters, 12 months, 12 summers, 12 winters, 12 years—a dozen doesn't seem to cover the cost. And not a solitary back-away shot of the place.

You gotta watch it around here—a week can last 25,000 words. A little summer work, a saga. And everyone has their rat stories.

For whom? It's what's up here that counts, and what's out there?

It's 1991, but it can't be. The book, started half a dozen years after, ends long before. Long before they're both dead, at 68. A book for the past, or for the future?

And when he dies, I'm not there, you understand? Or is it too much like work?

The Gulf War has just ended. Mike is there, he holds on, hasn't learned to let go. Can't, even now. But he's there. Three o'clock in the morning, he tells me about it, voice cracking. After months of late-night phone calls, cops on the other end of the line—my father driving around at three in the morning, blowing through greens. My brother has to get up out of bed, go down and pick him up, drive him home. Sick, he doesn't know what he's doing, where he is. This is how it begins to end. After months of this, after arriving the day after the anniversary of her death, Mike is in the apartment, once her apartment, only for a couple of minutes, then a couple of coughs, coughing up, blood gushed. And it's over. Over and done with, just like that.

And here we are, folks, here I am, nary a Depression soul left who would *think* of calling me Joe College: gainfully employed, broke, just a sign of the times, living on blues power, high-test, loving that dirty water. Meatball and potato man. A tribune?

Mike is there, he holds on, like I say. I imagine it now, as then, with urgency. Maybe you can see it. I must.

Dad, I'll call the hospital.

That look, Joe—you know that look?

Yeah.

Don't shit yourself.

Just don't shit yourself.

White, first-generation, veteran. Middle-aged, male, balding. A father, a finisher, a refinisher. By his own account, when sober, a no-good drunk.

By mine, he's at the end of his rope.

But he would do it all
   do it all again
     for you boys.
Still.

It's snowing again. Time to rub mink oil into my Herman Survivors. To write.
Still. I can recall it.
I can't.
Imagine her, as in a poem of mine, as in my troubled sleep—imagine her life ebbing in an emergency cardiac unit, imagine her dead, with earrings on, a year and a day prior. Imagine him seeing her, in the emergency room, table cranked up, almost to sitting position, death throes still across her brow, him having split off from her for so many years, or her from him, with her banded hand in his for so many years, so many years prior, on another continent, the two, the three the four, imagine what this means, from place to place, wage to wage, what it will mean, won't mean, can mean—

He's fallen, on rebound, for a woman who's experimenting with other women. Experimenting, or maybe just living her life. Something I don't understand as a teen. Nor does he. Her son Billy, from her first husband, a problem child.
He'll end up a dealer, she'll end up on crack.
Up in the Adirondacks, in a place called Storytown U.S.A., the eight-year-old will sit shrieking between Peachy and me, both of us holding onto him tightly, all three of us nearly tossed from an amusement park ride gone berserk.
Peachy's second husband—technically, she's still married, separated—a big guy, Karl Minski. Walks right through our front door late one night at 112 South Dolores Terrace. Twists the door's iron grillwork out of whack. I wake up, walk into the living room, pick up a vase. Or a lamp. I'm a chubby kid, wearing glasses, brush cut, in my underwear. Am I crying?
Please don't hurt my dad.
KARL—now take it easy! JOEY!—go back to bed.
I wouldn't have such a wimp for a son.
KARL!
My father is standing in his boxers. Peachy is screaming. My father has his left fist doubled, drawn back. He's a southpaw, boxed in the Army (I said this already, right?). Karl is drunk, looming.
Now just take it easy, Karl—

I hear them mumbling into the night as I try to fall back to sleep. They're in the kitchen now, all three. Talking. I can hear my heart pounding.

A while later, I hear Karl leave.

I've lost track of time, in my bedroom, in the dark, listening. The cops come. Troopers.

He breaks into your house again, you're in your rights to kill him. Pick up a shotgun, and blow him away.

I can feel my father's uncertainty there. Even then. The same uncertainty that's in me. He's seeing a woman who's not quite free of her prior marriage, something he punched Henry in the mouth for two years prior, after Dick Italia served Henry a subpoena as a correspondent in the divorce.

It's around this time that I forget to call Cognac back inside before I leave for school. He's our second miniature, silver-gray poodle—like our first, Bijou, my mother's idea. But she lets us keep him.

When my father gets home from work, no Cognac. He finds him along the side of Buckley Road—hit and run. When I get home, my father has placed him in the garage. His body is cold.

We have his body cremated, and bury him at the foot of the maple tree in our backyard.

A year later, my second year of high school, our first year at 501 Raphael Ave., I've lost thirty pounds.

To the best of my recollection. I'm not sure, might have missed a beat.

I was there, you tell me. Ten to one you *can*.

Big brother's guilt trip? Cop-out of the first person?

Not enough (applied) material for the reader to respond? Caught looking?

Flow interrupted, splashing on the tiles? Too reflexive, too self-conscious?

Mind games? Crawstuck?

Howling at the moon?

Broken feedback loop?

Broken feedback loop?

Bummer. Go for broke, you bet your sweet ass I will. No ends met, no means lived within, no shelter given. Raw, cooked, half-baked, like I give a shit—how hungry *are* you? Have you ever *tried* eating your words? Stick it in your ear. Up yours, fuck me, don't fuck with me. I didn't, couldn't say enough about Watergate, 'Nam—I was a spectator. So I didn't, couldn't, wouldn't. Mind your own business, your own (wordswordswords).[3] Pardon my French. And make it snappy?

Runthroughthe/suffragettebecausethehandmedown/mobile/idiot/thickas a/cantgetno/easytobe/feelin/morethansomethinhappenin/iwant/everybodys/ backover/ivelookedat/fromthe/upup/purple/sunshinemercymercy/go down/ifyouknowthaticould/oneway/papawasa/justlikea/backtothe/never thoughtthat/thelowspark/takeit/downon/allright/likeabridge/wordsbetween the lines/keepon/listeningto/whilemy/allthe/heymr/mybaby/itsgonnatake eight/simple/eveninthe/tuesday/war/heybabe/comesaturday/fallinginandout/ takemetothe/iknowa/heartsa/cantget/dontyouwant/whodoya/teach/theresa littleblack/mybest/closeto/cracklin/letthe/letsdothe/whosthat/movin/gonna usemy/onthethiniceofa/dirtywhite/ivenevermeta/J/smiling/hitchina/leanon/ ilove/ourloveis/whosmakinglove/canttyou/cantbuy/oneway/lawdy/ohhhhh oh/tin/revolution/skip the/dontyouknow/wegottaget, so sleep on it, and sue me—if you find it's too easy. Liberty ab(l)ated, piss-poor, liberties taken, not even a pot. Words meandering away from algorithm, back to the needy, to gunk, to engines of infernal compunction, or sleds, or AF of Labor, unlettered. In a groove on a B-side, hisses and tsks and stutters, between the sharps and flats, his and hers, theirs and mine.

PASS
WITH
CARE

But to know only bastard languages, bastard technologies, bastard dreams. M.M.M.S. wants *me*? The last man drafted into the Army enters a month after my eighteenth birthday, so why register? And the hell with work[ingatthe]shop, IUE Local 320, like my old man. Wasted, tearing ass, tail sliding with pickup, eurekas, sonic booms.

Who's finding other metaphors for reading? What's happening now to *me* will not be the same as what happened then to *them*? No matter what *we* say, no matter the nature of *our* congress? If *you* indict the man, must *you* indict the system, society, *yourself*? *I* write it down, *they* make it up? Yearly flocculation looms, a yeasty *us*?

Played, plagued with pronouns at will, and against?—*oneself's* fictive detection?

It's catchy, *she* tells me: *I make it snappy.*

Appearances to the contrary, I'm not here to entertain you.
Appearances to the contrary, I'm not here.
This chapter doesn't belong here.
This chapter doesn't belong.
Nor my venereal diseases of the mid-eighties. Nor, if you please, my sexual exploits of the nineties, warts and all. Nor any hint of the didact. But autodidact?

Writing does, and does not. Does not freaking exist to be put into a book.

As if it needed a reason, a look, a reprieve. A provider, provided for, provided—

"*Make* something of yourself." (See next chapter.)

Alfred Lubrano's got me pegged as a *straddler*. Smart guy. But I have five legs.

Listen: see, they came of age in a global theater that popularized the word "minority" as we currently use it. I read that someplace, as I read that the ceremony was on 7 September 1945 in Le Havre. Depending on your source, either four years after or one year before the first working computer (term used advisedly).

That age is ending. That word has a future. "Lest the firm peace and friendship . . . Pulled into, pulled out of. Anything but bogus, & chops aplenty.

About that time my relationship with Julie is ending. About that time. Seven-eight-nine years, ending. We weren't careful about birth control—we were lucky. And unlucky. ". . . now established should be interrupted by the misconduct of individuals—"

This could have been about us. More struggle, less progress.

How did I get here, there? Four presidents, dead and alive.

As time goes by, I expnce the world more and more through symbls—wrds, nmbrs, stp sgns. Symbols allow me to see patterns—relationships—more clearly. Perhaps they also permit me to triangulate further.

I'm unsure, sure. But out of the wordwork, *splinters*—

My mother stopped working when I was born, didn't start again until I was eight. At supper my father would talk about *the shop*, what a *sonofa-BITCH* it was, a mile and a half down Hopkins Road from Dolores Terrace. Sometimes he'd run the edge of a match-pack through his teeth. Bad teeth. Or bad gums.

Like me.

My credit as bad as his now, neither my life nor his in the words alone.

Sometimes, as here (or there), now (or then), it's the alphabet that creates the world I come to know. Sometimes it's numbers. Numbers or letters, when I sit down to work these days, *now*, all takes the form of steady dots of light on my computer screen. Before processed as words via computers, the letters consisted of seemingly less convertible, less mutable, more tangible scrawls of ink, or graphite. On paper, wood's first or second derivative—to speak metaphorically, as always.

Doubleclick, drag & drop, hickorydickorydock.

What you're holding in your hands at the moment—*then*—is a final draft whose earlier incarnations, many of them, were themselves of paper. And were, many of them, themselves of the screen.

Tales told unwittingly amid Unix, bar code, in vitro fertilization, Arpanet, chat. Divorced November '68, year of years, post "hyper-text," ending in genesis. Enter 501 Raphael Ave. in '69, a first lunar landing with artificial hearts and minds in mind, and a cold wind, northwesterly, over warm bodies made mostly of H_2O. Honors English and math in high school. Minor code-switching, n'est-ce pas? Bags, brags to britches. And these ligneous sentiments, would put it as Brautigan once put it, *We age in darkness like*—

Doubleclick, drag & drop, mouse ran up and down the clock. Persistent allusions to that Renaissance architect (draftsman), albeit harmony (balance) out of the question?

And what happened to my turquoise lederhosen with ivory fittings? Lost.

But loss or gain, it's all in our needs, and our desires. I've always had a good appetite. Some are less needy than others, some desire more. Why? Do you, will you have the time in your life to worry such questions?

Questions for this century, and the next. Like cloned calves and sheep, genetically engineered tomatoes, synthetic-DNA-based computers, neural-circuit computers, "paper" made of microencapsulated, electrically charged, image-altering cells, and printable cell networks—hard upon us, our nerve intact, autotelic.

Harder upon them.

We strike for pension plans, Joe. That's because we've got a lot of old-timers at the mill. Your union is full of young guys, always walking out for wage.

Yeah Stan, I know.

The Wojciechowskies live across the street from us on Dolores Terrace, one house down. Every other day, Mr. Wojciechowski is out there washing and drying his car in his driveway, an Olds 98. He's proud as hell of his Toro lawnmower. Of his lawn. Of his carpeted garage with wood-burning stove and TV—his family room. Mrs. Wojciechowski has a small business on the side, doing hair styling in their basement.

Mr. Wojciechowski explained to me once the difference between a trustee and a delegate. Mike was there, I think, we were all sitting on their front steps.

My father respects Mr. Wojciechowski, who's a bit younger. Not only are they both union workers—Mr. Wojciechowski has been working swing shift forever—but they share other values as well. They admire each other's successive Oldsmobiles, my father buying his last, an 88, when Mr. Wojciechowski picks up his 98. And they boast about their concrete, as opposed to asphalt, driveways. Concrete is something to which both men seem bonded. A sign of strength, permanence, it yet cracks under tension.

And if you walk down Hiawatha Boulevard—
The small houses, taverns near Crouse-Hinds, across from MacArthur Stadium, not far from the wastewater treatment plant, around the corner from Jim's Fish Fry. One world of work dwindling, hanging on, social returns diminished, majorities stuck in the middle—again—all in the midst of Metcalf's "manufacture of rock"—
LBJ is dead, long live LBJ?

Woodworking is much like poetry, only with thicker paper.—ad for diynet.com 5.11.03.
My father and Stan Jr.'s father will both die before they hit seventy. What's worth seventy years? What's four years worth?
A new lawn mower?
Stan will land a job at the steel mill, like his father.
Lineages linking and unlinking in the dim light: me, lines of work batched, broken. Botched? Without delegation, nothing entrusted, without authority, nobody responsible. Give as good as you get—but how could, could he wish this for us?
Shoes filled, unfulfilled. In the Salt City, no less. Where Carrier Corporation becomes United Technologies. Crucible Steel, Colt Industries. Where Rockwell vanishes, or seems to, as General Electric shrinks to GE, only to be displaced by Lockheed Martin. Where GM circle loops lighter and lighter lunch traffic around and around, past more and more acres of silent, weed-cracked asphalt—acres dedicated to absent histories, secured against intruders by chain-link fence.

Summer 1978: my brother contracts mono. Bad.

Hearing voices? Lying on the couch . . .
Catawba, Isabella, East Division . . . Oswego Boulevard.
Ma and Pa's old house on Oswego Boulevard. Remember that, Joey?
I remember, Dad. Kinda. I mean—I was there only once, right?
And that piano I made Suzie, the one that plays "Beautiful Lady," now in pieces. In pieces.
  lying on the couch
  on the couch, he's lying on the couch
  what was that song John Wayne sang?

What's that Dad?
QUIET! He was whistling—something about a lady—in yellow—

"What good has readin' and writin' done *you*? Look at ya—in an apron!"
(Vera Miles as Hallie to James Stewart as Ranse in *The Man Who Shot Liberty Valance*.)
A photo of my father peeling an eggplant, head bent over, bald spot nearly the same size as the skinless fruit. No apron.

Bloodstains on the rug, 1991. The Gulf War is just ending, the apartment is almost emptied out, a familiar pained face babbling sad-m, sad-m, sad-m . . . . Soon I'll be on my way, back to Illinois. Food and drugs were administered, time shared. A few pieces of mind were given, but peace of mind received? Question marks sufficiently handed out? The world united, the press satisfied? My first meatballs big as snowballs.

Some guys, Joe, you have to kill 'em, if you get started with 'em—
If a group of guys come at you, you pick the big mouth out of the group. You flatten him, the rest'll scatter—
You gotta be dirtier than they are, meaner—
Raised amid lake affect, snow on the old screen. Dirty snowbanks, twelve feet high, snow jobs, blow jobs, no jobs. And what the hell ever happened to Tarby's, in the village? Those cheeseburgers. Ma & Pa's meatballs?—today it's high-end chains, a palette we never knew. General Electric-s, general electrons, general elections?
He—a drunk or no, the obligation was met. He always voted. Even plowed.

Takes guts, to a man, to a woman. Like when the bunch of us were sitting on the church steps at Dolores Terrace and stood by while Bob O'Day commenced to kicking the shit out of Steve Alonzo. Alonzo took a real shellacking that day.
So we were all "friends," so what. It makes no matter that we made a few half-hearted appeals. Bob had become a fuckhead by then, like his older brother Peter.
We didn't have any guts, not that time around. Not the bunch of us.
With my father—something in him would click, just like that. So it seemed. Past a certain point, he lacked my—any vocabulary for compromise, for backing down.
Even for the wrong cause. And in the wrong—what then?

* * *

Bluster? If I learn anything from him about fisticuffs, it's never to telegraph a punch. What it does to his metabolism to swell his soul to split-second rage, who can say?

A tell-all? Family skeletons? Rumors?
What of it. Oh yeah: and writing is no boxing match.

Robert De Niro's father—Robert De Niro, Sr., a poet, sculptor, and most notably a painter, both expressionist and abstract, who studied under Joseph Albers at Black Mountain College in 1940—was born in Syracuse the same year as my father. He died two years after my father. Did each man's eyes hold similar intensities of portent?

Did I mention he used to sing? Around the house, back when we were "youngsters." And rarely, on Raphael Ave. Not a bad voice. I still have his *Song Hits* and *Hit Parade* magazines, the earliest dating back to the early forties, when he was in his late teens, looking anxiously at the draft. Like most teens.
Every now and then during the late sixties, when he's in his late forties, he'll ask to read a comic book of mine. I'll sell most of them in the early nineties, when I'm in my late thirties. Now in my early fifties, I find in works like *Maus* the prodigious flowering of comic, comedic haunts.
Would he be interested? I know my mother would.

So what.

Where's Mike?
He's there, even if he's not.
But like I say, I can feel my old man's uncertainty here. Even now. The same uncertainty that's in me.
This is my place, this is where he left me.
And she, she the stuff to keep going?

Sleeping, I'm awakened by loud voices. My parents, arguing again. I put my hands over my ears, but I can still hear them. Hearing voices.
Go on—you don't know what you're saying.
I don't, huh Suzie?
No you don't. You're being ridiculous.
Really? Are you gonna tell me, Sue, you don't know Henry Poster?
I work with the guy, that's all.

I was there last night when you got out of work.
What do you mean?
I mean I was there—I saw you.
A silence. I can hear it.
What are you doing now—spying on me Joe? Is that it?
This is for twenty-two years, you no-good two-timing TRAMP!
I hear a loud snap. Almost a snap. And my mother's groan.
Then a long pause. I'm listening, wide awake.
Or am I?
The voices last for years.

He'd almost died once before, of appendicitis, circa 1940. As a kid, I see him twisting and turning on my parents' bed, groaning, suffering from kidney stones. The only time I see him in pain, until the divorce.

I went head to head with him, but could I handle him? He knew how to handle himself, heel to toe. Have I somehow walked there, in those footsteps?

And forgiveness?
What for?
No harm done?

Don't shit yourself, either of you.
Both of you.

I've never been this poor in all my life.
Not even in the thirties.
If it weren't for you boys.
If it weren't for you—
If it weren't—

Henry doesn't seem such a bad guy. But too friendly, too eager to please. When my mother introduces Mike and me to him, we're not sure how to act, react.

My father talks to himself, late at night. I imagine I hear him.
I can recall, he says, sitting on that back step. It's the back step to the back door of our garage, opening into our large backyard. Large for a kid, for six kids. At 112 South Dolores Terrace. It's cold out.
I'm stark naked. Drinking a beer. He says.
My wife has just told me she wants a divorce. She's just told me she wants a divorce. My wife.

What's the matter with Dad?
Never mind your father—he's crazy.
Emphasis on cray-zy, with her French-German accent. A world inflected.
Did I say I voted for Ford? Me, a lifelong Dem, hoping to stave off GOP—
backlash? Flashback to '68. Yougottabefuckinshittinme.

A dream, another one, in even numbers. When I awake, Mike is walking in his sleep. Then talking in his sleep.
You boys will be scarred, he says. She ruined her love for you. You won't respect her.
She's left scars:
You'll love her, but you won't respect her.
Scars.
Nothing is a dream, it's all a dream.
Little Gem Diner now, 118 Maltbie Street then.
You were puffed up like a frog.
Peachy's post-coital remark not intended for my ears. I'm in the next room, sleeping. Trying to.
Try to.
He was in his late forties. I can imagine it now, almost—almost there.

Susan B. Anthony was born in Adams, Massachusetts, in 1820, one of seven children. Her father, a Quaker Abolitionist, raised his sons and daughters alike to be independent thinkers, and Anthony could read and write by the age of three. When one of Anthony's grade-school teachers refused to teach her and her female classmates long division on the grounds that this would never be of use to a woman in her housekeeping duties, Anthony's father removed her from the school and began homeschooling.
In 1852, in Syracuse, Anthony attended her first women's rights convention.
First minted in 1979, the Susan B. Anthony dollar coin is about the size of a quarter, causing some public consternation. For this reason, and because of general discontent with the Carter administration, the coin is referred to derisively as the "Carter quarter."

Helen is a good woman, trying to raise her daughters Nancy and Jane on her own, in a small house just the other side of 501 Raphael Ave. She's from the country, introduces my father to country-western music.

He's going to punch Helen in the mouth, too, a few years later. Punch pulled, but hard enough to swell her jaw.

Years later, he hides from us the fact that he's been seeing Jane, Helen's youngest.

I don't understand, don't want to, must. Disgusted, don't have to tell. Have to, have to tell you, hate to, whoever you are. Setting too fast though—needs retarder.

Let it out, let it all hang out, out with it all. Out.

Fast forward to—to when, exactly?
Mike has moved out. A second time.

Your ass sucks canal water, Joe.
So does yours, jerk.
Regional? We laugh. We always laugh, somehow, together, at the end. We watch lots of movies together, on TV. We both love the Mills Brothers, Nat King Cole.

4.29.00: *Jeopardy* category "City Anagrams."
Answer: "Say Curse, New York."

Dad, what's this song again?
I'm holding a banged-up, miniature replica of a grand piano, blond finish. I've got the top open, and the aging music box inside is tinking out a melody. My father makes a number of these as gifts—the first one, the one I'm holding, for my mother, a jewelry box for her birthday. Years later, after he's gotten the details just so, he crafts his finest piano for his divorce attorney, Mr. Hall. Another gift, Mr. Hall having treated him, as he sees it, fair and square. A smaller piano, without a music box, sits on my rolltop.

You know I made that for your mother. Before she lost her mind.
Yeah, I know. That's why you broke it too, right?
She lost her mind!
Yeah, OK. What's the name of this song again?
Beautiful Lady.

Beautiful.

Desolation.

Desolation.

Beautiful desolation.

A *muckraking work of*   *memories of*   *myself watching*   *my father watching*

Someplace on the dark side of the moon, the moon is dark. The street is empty, the crickets are chirping. Less and less traffic around here these days. The Thruway is still, is always busy. WOUR out of Utica is not what it used to be. None of us will be, nothing is. Later than it seems, or too late? Can't be, mustn't be. *Later—*

Dear Mr. Amato:

We appreciate your realist convictions, we really do. But the truth is, you're just not *literary* enough.

Very Best Regards,

From a dream. "Come and hear how this touring author turned personal tragedy into a bestseller." That was much, much later, on a postcard.

She's right. I can appreciate the marketing ... considerations. But what's the distinction, intrinsically speaking (if there *is* one)? And reception will be altered *how?*

Read all about it: Author shoots from the author's hip. Very *uncool* (1953).

Not innocence, not bliss. Outsides, but no outlet.

OK, I've been fibbing—one maybe, but no shortage of ignorance. Whadayasay?

My brother has moved out again, a second time.

His former bedroom is full of junk. Rain is coming in through the roof, over the bedroom window, dripping across the walls. And rain is dripping down over the light socket in the center of the ceiling. I disconnect the wiring, in doing so get zapped a bit.

* * *

Angry, wanting love, I could write about anger, I could write about love. About how each may require the other, over a lifetime. I could write, they could write their names. He could write, but he didn't. It wasn't his way. He told stories, as did she. They wrote to each other, across the Atlantic, in 1948. I have their letters. Love letters. I struggle then, like now, with these stories. Just stories.

This is not self-privation. We weren't mountain men, nobody was named Ethan, we were all registered voters, we didn't have much of a say in it. A more open approach to farting—what of it? Go fuck yourself with your—self-reliance?—

—*schadenfreude*? Go f-u-c-k yrslf. Better, try to write it out, w/o amendment. Out of your system, without a say. Without a hope, on a wing and a prayer. Beloved?

To undo the undone, one unmakes the maker. Unfinishes, refinished, unfini period.

What was he afraid of? Of death?

Don't shit yourself.

We're watching a documentary on TV about Korea. His body is ailing, he's developed a slight lump on his head now, a year after the chemo and radiation. He's lying on the couch, telling me about the war during a commercial. The documentary continues, covers the Truman-MacArthur brouhaha. Cut to another commercial. He begins to speak, slowly.

Old soldiers never die . . . they just—FUCK OFF!

I laugh. Lurking beneath the pain is that old irreverent self, that completely lucid fire.

Yknow Joe, if they'd ever drafted me to go to Korea, I might have headed to Canada. I talked about it at the time with Suzie, we agreed. No way was I going to find myself in another war . . . .

As much as he enjoys war films, he sees through the propaganda of self-sacrificing patriotism. And yet he understands the necessities of war, once waged. He understands sacrifice.

Me, I've never been shot or shot at, stabbed, or abused. Hardly been punched. As an undergrad, never held a job during the semester. Making so much of so little then, OK. Your typical scrivener.

They concur: a speck of color in a grey landscape draws too much attention to itself.

So let's not get ahead of ourselves, they say. Let's be realistic.

* * *

Money, it's an assumption. Assumption, it's a cemetery. Four bodies I knew.

So what was he afraid of?

Of the consequences of dying? Repercussions?

Or of death's eternity, the eternity of a moment?

Without amendment: he was afraid—I could feel it at times, and I'm almost afraid to say so. He had one of the strongest constitutions I've ever seen, but I'm sure of it—anxious, and afraid. Even defiantly so.

Afraid, lying on the couch. Fighting, lying on the couch.

Don't shit yourself, get up and go to the bathroom.

Of course he did, he always made it. And then a report on how it went. The body takes over, any body.

Do you want me to draw you a free body diagram? There are no free bodies, never were.

I'm thinking to myself, no rest for the wicked. Where did I first hear that?

—there was this pretty black woman I knew once, she was so nice—

Huh?

—this black woman. Very pretty, well mannered. She was the nicest person, Joe.

That why you don't like blacks?

—shut your mouth!

Not the best timing. Never was a best time, never would be, can never be the worst of times.

If it weren't for you boys.

It's as though we aren't even there.

An old tube chassis, lying there. No replacements in hand. Or four-bolt mains, not another short block in sight. Addicted to work, chemicals, pain, beauty?

I'm asthmatic, an asthmatic. My grampa had emphysema, most likely.

It's as though I'm not even there, here. As though you're not here, there. Where is everybody?

Gone, going, keeping on keeping—

I think I understand, you understand?

A knowledge of simply being alone. Alone, and unamended. Nothing in writing, no will. No witness, nothing. Nothing but tools.

Alone, as was she.

* * *

No, of course it's not fair. I regret it. I want to do it, I want to do it all, again.

I don't regret it.

She nearly married a Frenchman, Bernard, a man prone to seizures. The seizures scared her away.

Aunt Ilse tells me this, decades later. She lives in Toronto, alone now that Eric has died. He was an artist's son, my uncle Eric. Though educated in English schools as a child, a Frenchman through and through. As is Eric's son, Dan, who lives in Toronto with his Quebeçois wife Thérèse and their two kids, Serge and Michelle.

On this side of my family, my mother's side, is another history, or set of histories, punctuated by money. The family owned a restaurant, a house in Landau, even a piano (which my oma played). They were decidedly bourgeois, until the war, when they lost their land. Up in the attic at 501 Raphael Ave., I discover four old serviettes, each bearing that embroidered B for Bourgoin—my mother's birth name.

I can recall my uncle ushering me to bed in the wee hours of a Christmas morning. My aunt and uncle are staying with us at 112 South Dolores Terrace, and I get up to play with one of my gifts—a sparkle-color paint set, as precious to me as my moon rocks.

Jo-Jo, c'mon, let's get to bed—you can play with that after a good night's sleep.

His J is just a bit soft, just like my mother's. I have some of Eric's paintings, oil and watercolor. He managed a paper warehouse in Toronto for years—Hilroy, for years he gives us Hilroy writing pads, notebooks.

It's my aunt who'll explain to me, years after Eric is gone, and after both of my folks are gone, that her actual paternal grandfather was Jewish, and that this man never married her grandmother. Her grandmother, already with child, married another man, the family hiding its Jewish heritage during the war. And after the war, the secret lingered on, far as I can tell, due to prejudice.

Most years I'd write the Christmas cards for the three of us, signed "The Amato Family" or "Joe, Mike, and Joe Jr." to the folks at Dolores Terrace.

Joe, make sure you send one to my brother Dominick. And Dick Italia.

OK Dad.

Mike would wrap the present for my mother. Perfume. A new kitchen gadget.

* * *

I have my mother's postcards and letters to us, after the divorce, and a few old photos, taken at a beach someplace in France before my mother met my father. And I have two of my mother's leather book jackets, with her initials.

My mother, childhood spent in the Alsatian region of France and Germany, will always love to hang wash on the line. I think of her carefully picking the Japanese beetles out of her rosebushes, clipping her chives, watering her tomato plants along the side of the house. I think of her, smiling. She likes to shop at Lerner's.

She returned to France in 1948, almost left my father then. But her mother would hear nothing of it.

In fact I have all of my family's belongings, all that will ever and can ever belong to me.

It's not much, a world.

Must we all? Must they, we, be alone, in the end? And the ending?

And to amend?

The DePasquales ("of Easter"). On my father's mother's side, an uncle ("Tom") who sails to Sydney and an aunt (Concetta) who sails to Argentina, never to meet their nephews. My gramma's older brother—Francesco, a fisherman and best friend of my grampa's, with him in the Italian Navy—drowns in a boating accident near Boston. His kid brother Antonio marries the widowed wife, which wife my gramma will henceforth shun as the woman who stole her brothers. Uncle Tony's daughter is my father's cousin, the upper-crust Francesca, his son Jumpin Joe D. Jumpin Joe is a second-story man, like my Uncle Sam. Partners in crime, literally. Joe got his name from his jumping ability.

Sam (Salvatore) is the Amato family disgrace. He makes the front page of the *Post-Standard* in the late forties, a notorious burglar. My folks talk about him in hushes, my mother tells her sister that my father has another brother named Jeremy.

I should check it out sometime, see if it's like they say. No room for my uncle Sam's life story here, in this story. I picture him as—not a made man, no, but self-made, to the extent the humanmade self may be so autonomously identified. My folks never talk of Mafia connections, nobody does. Maybe he knew some people.

One thing is certain: as he grows older, he mellows out, wises up. My uncle Sam—you can tell he's trying to make up for his past. He's meant well ever since he got out of the can, and his father died. Delivered pizzas in his early seventies, delivers for charities now. He's had a couple of hip replacements, and a stroke—can't speak too well. And most recently, has lost that leg. The one with the bullet?

After we're married, Kass meets Sam. They get along well, play cards together. That sonofabitch *cheats*, reports Kass, smiling.

Sam lives in Tampa in a trailer park, with his long-time girlfriend, Maryetta. My aunt. He couldn't take the cold any longer.

I should see if it's like they say there, I should check it out sometime.

What sort of writer am I, have I become? Check *what* out?

Sam is dead. Maryetta, in her late seventies, is alone in that trailer now, trying to find work to make ends meet. Dominick sends her money.

Frank is gone now too, my Aunt Mary alone in her little house in Eastwood, trying to hang on.

My father hates *lazy shiftless bastards*, as he calls them.

Some people just don't wanna work, Joe.

I wonder why this is, search for answers in our social system. My father could himself have taken a job at minimum wage—don't think for a second I don't know this.

My mother works her ass off. Her job as a receptionist puts her in the company of people who have more formal education—she greets "important people," answers the phone, types, plans, arranges. Has to keep up *her* appearance. When she's mistreated, she sticks two middle fingers up under her desk, out of sight.

Mike and I own two decades' worth of GE-emblazoned writing pads, notebooks.

When she dies, after a few years of sporadic and undiagnosable flu-like bouts that last several days each, I find in her bedroom, addressed to Mike and me, a letter detailing her finances.

My mother, she's lucky at poker, pinochle. "Suzie you louse!" shrieks out of the kitchen whenever the Squads come over for the occasional Friday night card game. Squads: my grampa's sister Rose marries Frank Squadrito. My father likes some of the Squads, as he calls them, doesn't like others, in part because of a rent squabble years prior. As a kid, I like his cousins—yappy Frances and friendly Santa, they're funny, kind ladies. I'm in awe of Frances's terraced backyard, and Mike and I like Santa's son, Alfred. When we visit Santa's family at their house in the village, Alfred and Mike and I talk monster cards, and tune Alfred's shortwave radio to pick up signals from distant lands, staticky fragments in strange tongues. We listen together in the dark bedroom, imagining the shortwave police closing in. On one occasion, the older folks call us in to watch the Beatles' debut on Ed Sullivan. Santa and Fritz have an older son in 'Nam. Struggling to recall those small pictures—paintings—of my mother's, tiny vistas of small towns in Europe. In small frames. I've lived there, I think, I've loved there.

Body work coming, bundle up. And sink money into? Plan to leave what? For whom? She could bend down in a field of clover and pick out the lucky ones.

They all show up at the funeral home, Italian surname after given name. For my grampa, for my father. For my father, Helen is there. Does she know about her daughter and my father? She's getting older.

Mike and Joe, your father was a good man.

Are we nodding? We hug, kiss Helen.

Peachy never shows, Johnny Palamino is there.

An obese woman, Shelly, whom my father has befriended from the tavern down the street, has brought her daughter along. Shelly is breaking down into sobs and wails every so often.

We all try to ignore it.

A week later Shelly will try to convince my brother that my father bequeathed her his Chevy before he died. Mike won't buy it, and the title is in his name anyway.

And—what's his name? Little guy my father's age, gray, friendly, nervous, talks fast. Jimmy—Jimmy Camiso I think it was. I think. He was there too.

Your father he was always smoking your father. Joe always had a cigarette in his hand your father did Joe did.

Mike and I thank him for coming, shake his shaking hand.

So many of these people don't have long themselves.

Cold cuts during card games—good hot capicola—and small change sprinkled on the floor the next morning for Mike and me to hunt down.

Curbside now, chickenshit car repairs in Chicago, the city that works. I'm still at it, trying now to keep up another shitbox, my mother's old Escort—keep it alive, move it through its second hundred thousand miles. Part of my inheritance from her—she left Mike and me thirty grand apiece, all said and done. Knew how to save for us.

Real half-assed jobs—I've lost my touch for it. My hunger.

Graffiti language in the offing now, "pilots" at the end of last chapter's numerical palms, qualms. *Hey—what are you looking at?* Pedestrian traffic here can be heavy. If you head upstairs for ten minutes and leave a screwdriver on the street, it walks. The city has outlawed the sale of spray paint in cans, hoping to cut down on tagging, but has not outlawed their use. Air quality aside, *tagging?*—but that Postal Service mailbox down on the corner—been tagged four times, twice just a few days after a fresh coat of blue. So: I drive outside city limits, come back in and set to work, lakeside, lopping on mat & resin like cream cheese. Then a couple of coats of enamel, that familiar hiss.

Half-assed job or no, I'm almost proud of myself. Almost.

My fingers swell as I work, I'm mumbling, my wedding band tightening.

No kids. Kass and I have tried, time has run its course, has run out. Never quite certain it's for us anyway. Still.

Kass never met my folks, they never met her. We met late, but it couldn't be helped—some things take time. It's in Illinois that I first learned of Syracuse Cultural Workers.

My mother's services are graveside. She's cremated, as she wished. Her friends from Schenectady are shocked, grieve from a distance. The CR & D newsletter runs a tribute column, unheard of for staff workers.

The snow hits the ground, refusing to melt.

Refuses to melt.

I wasn't there, it couldn't be helped.

I was there, I wasn't.

I was.

I'm not here to tell these stories. But to mystify these homemade myths? I went through all her things, sifting, saving, tossing away, gifting. Mrs. Becker helped me.

Note to self: remember to ask certain friends (*friends*) why we stopped talking.

My oma died when I was three. She's buried in Toronto—Weston, to be exact—not far from where my aunt, Ilse, lives. Not far from the final resting place of my uncle Eric, from Ilse's future resting place. I never knew my mother's father, who died five years before I was born, in France.

I couldn't make it to my gramma's funeral. Living then in the city that works, I was working. And she was so old. She was in her own Old World, perhaps more Siracusa, that port in Sicily of ancient Greek beginnings, by most accounts the birthplace of rhetoric as a technique, and an art—than a town on the Erie Canal.

Three centuries ago they were gathering snow from Mount Etna, storing it in caves, compressing it, bundling it in burlap and straw for transportation to Rome and beyond. File it under "The Invention of Gelato, Arabian|Sicilian, 9th century."

Ever working? I chose not to stop working. Still, I could have stopped, driven back. But my brother would be there. This could have been about us. More struggle.

I was there, so you tell *me* what happened. You tell me. It takes time.

I tend to amplify. Take those waterfalls upstate. Comfortably private places during our teens, yet strangely foreign. I think now they went public too fast, overcrowded by the public urge to find something to *do*.

Lots of ice cream joints too. Even today. You know those Italians (above).

Mike's longtime partner Peggy informs us that the oversized mall on Onondaga Lake's south shore—the mall that bumps up against the city's wastewater treatment plant—has become a notorious site for suicides. Peggy's mother keeps her up to date on jumpers. They stroll up to the atrium level in the mall, and over the railing they go.

What is it makes us—lets us chuckle over this? The faint munching sounds rising up from the food court? The merry-go-round melody? An overbaked Greek tragedy.

Or a slapstick of the dispossessed? OnonDada? Webster's Landing gone awry?

Acorns and hickory nuts were scattered around the old neighborhood, potential projectiles for our assaults on squirrels and stop signs. I think now we didn't look up enough, didn't seek out conspiracies of source and renewal.

I remember one time. Mike is there, back from Rochester for good and about to move out for good, and my father is there. And I'm shouting at my father. The reason is trivial—I can't recall what it was—and I'm head of household.

It upsets my father so much that he rushes into the game room, grabs a shoebox, starts throwing a few things in it. Says he's going to leave. Just up and leave.

I put my arm around him.

Dad, look—we love you.

That's the only time this ever happened—this direct appeal.

But I tend to amplify. And if this direct appeal is a singular breaking point, or a singular sign of healing—if this alters me, my capacity to express myself, push come to shove—it doesn't change the dynamic any. Not in the long run.

And me putting my arm around him: we weren't raised to hug men, even one another. Handshakes are the rule, or friendly wrestling. My mother is the touchy-feely exception, her final departure bringing Mike and me to our senses some.

I remember another time, though, toward the end of our lifetime's stay—clearly. Or it seems clear, now. At the time, I'm not sure—still, I lose my temper. And this event helps change the dynamic, all right—brings me down to earth, at his *and* my expense. Helps me to see—see the unconditional momentarily conditioned, the price of persistence, of endless exposition, that failure to show, all told, even in words—*not love, no*.

Anthony's final words before the public: "Failure is impossible." She will die in Rochester fourteen years before full ratification of the Nineteenth Amendment.

We fail all the same.

Mike has moved out now for good, and my stuff has spread all over his bedroom floor. A weight set, bench press, and boxes. Yellowing comic books. Bike magazines. I've begun to collect the final page-long essays in *Time*, trying to establish a body of knowledge of my own doing.

And a box that at one time held three bottles of German Riesling. Inside the box, what I believe at the time constitutes an investment: 100 shiny new Susan B. Anthony silver dollars, issued not long prior. We call them silver, but the truth is they're not silver, exactly—dollar coins, some small percentage of which is silver.

Silver or gold mingling, what's money? And how to mint a man, a woman? You need brass, you need iron to get by, to see things through. Purse of one's own.

Like my father, I've never been too good with money. But I'm paying all the bills now at 501 Raphael Ave., so I figure I may as well try to plan for the future in my own small way.

So my mother is in town, visiting. She stops by 501 Raphael Ave. first, then the two of us drive up to Swallow Path, where Mike is now living with Stan, Greg, and Dan. Four guys in a four-bedroom house in the burbs. And three motorcycles parked in the driveway. Mike and Dan in particular have a fondness for crashing Frisbees into parked cars. The neighbors seem a bit concerned. Once in a while we'll spot one spying at us out of a bay window.

Anyway, my mother first stops by 501 Raphael Ave., like I say. Shoots the breeze for a while with my father, and with my father and me.

On this particular occasion, I tell her about the Susie Bs in the box.

Oh! That's a good idea, Joey!

She's always a little too enthusiastic about my ideas.

Here, lemme get 'em.

My father is suddenly silent. I get up off the couch, walk into what is now the spare bedroom. I find the wine box underneath some magazines, grab it.

The box is empty.

Something in me snaps at that moment. Something having to do not simply with how ungrateful I find this, but with how childish it strikes me, how trivializing. Almost as if I had now to deal with yet another problem—the adolescent mentality of our household. Which I'm just now beginning to understand as hurtful in so many ways, however requisite to coping it's been for the three of us. Plain hurtful.

As though, relative to one another, we've never grown up. And me, 25 and stuck here now with what the state now calls my *dependent*, not knowing how to leave, not knowing how to say what needs to be said. What needs to be said not about staying, or leaving. What needs to be said about *needs*—

Something in me snaps at that moment.

Holding the box in one hand with the lid open, I stomp into the living room, where my father is seated on the couch beside my mother, the two talking, engaging. But my father is a bit tense.

Where the FUCK are the coins?

I shove the box in his face.

He pauses.

I needed the extra money, Joe.

For what?!

Beer. Smokes.

Listen, you fucking JERK—if you want some money why didn't you just FUCKING ask me? What the FUCK is wrong with you you fucking JERK!

No answer.

What have to you got to say for yourself, FUCKhead?

No answer. My father is looking away, taking a drag on his Parliament, his hands unsteady.

My mother is making that sound with her tongue, shaking her head. She looks at me and stands up.

C'mon Mom!

I stride out of the house with my mother, leaving my father to sit there on the couch, alone.

Alone.

* * *

A mile down the road I'm already feeling like a stupid asshole. But I'm good and pissed. My mother is still shaking her head, talks to herself and to me at the same time. We're both conversational thinkers.

What's the *matter* with that man?—sometimes there's just no talking to him.

Jerk! What the fuck—why couldn't he just ASK me for the money!

Joey, I told you—your father is a pisser. He'll give you the shirt off his back. But when it comes to money, he just can't be trusted—never could.

Mom, this is not the way he's always been.

You don't know your father like I do.

Mom—

Still, Joey—it's not right, you shouldn't talk to him like that. It's not right—he's your father.

I stop talking.

What do I feel—then? What do you feel? After all, what *can* you feel?

Especially when there's nothing stable around here, not even that final vowel.

Did I say Carrier Corporation? In October 2003, they'll announce plans to move overseas—Asia—and lay off 1,200 workers. That's a loss of 1,200 x $15/hr. on the average. One million manufacturing jobs will leave this land of manufacturing over three decades.

Loss, losing, losing self-control: a way to control when you have no control. Some would say, when you're low on self-esteem. Low on hope, without a say.

Not the way of Mediterraneans with money, power, confidence. They sit silently by, wait for their chance. Don't wish to tip their hand, and talking is a tip-off because conflict for them = argument, = stick it between your teeth. *Omertà*, OK, but— =	when my father, he loses control—his anger controls you. Same goes for me. So I lose it—with him. N↑ True?	And here—I lose it around here, I lose you. But I don't wish to control, am hoping more for participation? So how not to control, and how not to lose it? How to distribute fairly, fair shares of comprehension and understanding and responsibility?
Stop looking for it? we are ourselves here sadly subject to the -type. of	a reader-writer conspiracy of	There's talk these days of a revitalized Salt City. Of the Salt City as an east-coast Silicon Valley. Silicon—the second most abundant element of the earth's crust, occurring naturally only in combination with other elements, as in silica or silicates; the latter, chemically, the salts of silicic acids. Silicate as in clay. el 407' lat 43.1 N lon 76.1 W clay
Stop looking for it? 3rdGiuseppe carpenter? 1stSalvatore savior? 2ndFrancesco free? Purple ♥ 4thDomenico on Sunday? *Get up in the morning and* did somebody say *demonstrative*?	silence? arc hive of mull-in-yahn? oy. mameluke? Oi. Fosdick —> an obscure reference to working title Syracuse: "the parts near Surako" (a marsh) John Wilkinson names it—1819 "For her anti-slavery work Anthony was hanged in effigy crowds jeering at her image on the streets of Syracuse."	I'll have an argument with my father, years later, about going to the hospital. My mother will be present then. He'll refuse to talk about it, get up, pick up his lighter and smokes and try to walk out of his apartment. And I'll grab a hold of him in the stairwell by his left arthritic wrist—that same wrist that used to hold that mighty left paw of his over his head, threatening to wallop us if we misbehaved—I'll hold onto that wrist of his tight to prevent him from leaving, to get him to hold on, to hold onto him. And he'll wince.
He'll wince. And I'll let go.	No matter that he'll phone me later that night to tell me that he's reconsidered, that he'll go with me to the hospital after all. No matter that the treatments will give him only an extra year. (What's a year worth?) No matter that I'd do it all, do it all again, for him.	No matter, because I'll have learned that, not wanting to, I'll have hurt him. Like he's hurt me, my old man, whose life I'll let go of even as I hold on. Whose life I'll take a long hard look at, over and over.
Mafia—> lawlessness <—*mahyah* (Arabic), "boasting" exaggeration? "our thing" alkali metal—>*al-quili* (Arabic) <—"ashes of saltwort" —>basic (pH>7)	I'm already there—my short life, so long. e.g., Na, K [see Manindra Agrawal]	And looked at over and over, his life, my life, any life: not even a friendly amendment, no, not a chance, not entitled so. In the end, in the final manuscript, at the last call, when our number comes up—what is left to amend will be lost on us.

11.
Notes toward a Supreme Fiction

> And all this science I don't understand
> It's just my job five days a week
> —Elton John, "Rocket Man"

AUGUST 1969. It happened before. Plat become plot, weeds to a second story. No use explaining, no use easing into. Outsides, but no outlet.

We take the turn, the three of us. The turn winds us around, almost in a circle. A few hundred yards to the tavern at the corner, and we turn right. You can just see it—a run-down two-flat, maybe fifty years old. Siding pasty green and peeling, looks like cardboard when you first lay eyes on it. An old man up on the porch, waving us around back—or front now, where you park and enter. Five yards of grass and weed below him, running along the west side of the house—its entrance, or what used to be. Below and across from him, a wooded lot, and running between this lot and the house a short access road that ends abruptly at a fence gate. This access road is Raphael Ave. It dead-ends a single block to the north at the New York State Thruway, and picks up again on the other side. We found this out the hard way.

We're driving down Cambridge Ave., the perpendicular to Raphael Ave., a road that dead-ends at the back of the house. To our left, across from the north side of the house, a large open field. To our right, you can catch a glimpse of large rusting tanks just the other side of a fence that runs along the south side of the house, dividing the lot the house stands on from a storage yard. East of the house and past where you park, the creek.

The neighborhood itself a small parcel of a baker's dozen houses, contact with civilization to the north cut-off by Thruway construction. Constant east-west migrations serve as a sign of moorings gained, and given up.

We pull up. We're driving a 1960 DeSoto. Push-button automatic. My father already wants to leave. I tell him we've come this far.

* * *

28 March 1979. *Now I am terrified at the Earth.*

The date serves primarily as a bookmark.

It happens over hundreds of thousands of years. The species that talks with its hands, talks with its eyes, talks with its mouth struggles, in numbers and singly, struggles against geological night. The limits of the physical universe dictate to all creatures, all perishable things, all things, fixing the world in time, timing its events.

The engineer's wish is to see things built, to work out the details of the as-built world. Unwritten practices of planning, writing, making, moody and methodical and perpetually in flux. Duration bears down on telos, making of perfection a state of mind. And minds change. What we find in the details, search as we may, are props for thought, of thought—deities don't bother with such stuff. In a production plant, you enter anew each Monday morning already on the job, wary of oversight, resuming

in medias res. And that, as they say, ain't the half of it. The professional-organizational pecking order tends to stipulate a least common denominator.

I need that job, that paycheck. We need it. And where there's need, there's urgency—livelihoods are at stake, and, through that curious but constant relation between earning and living, *lives.*

One purpose of my education—the sole purpose my father really understands—is to help me wrest control of my life, in ways he himself hasn't managed. But will my lot be that of overseer, swearing off the horse sense of seers?

Engineering, as a practical science, is all about control. You learn that thinking this way, not that, produces tried and true results. You learn that time-tested techniques can have, have had, will have powerful consequences. The long history of design failures and successes has altered, adjusted these techniques, sometimes radically. Over time, technique itself becomes a habit of thought, a set of procedures. And as an intern in the field, you learn over time to adapt to the field's designs on *you.*

Design: not Moholy-Nagy's "design for life," no—nothing nearly so self-conscious of its systemic reach, nothing nearly so steeped in critical contemplation, the nuances of social thought and educational planning. Engineers learn design today in terms of programmatic methods, methods that produce predictable, optimal outcomes. In reality, the design process is contingent upon so many variables—like the weather. And just as contingency is the axiomatic exception in weather forecasting, the engineer's emphasis is on what *can* be predicted. A makeshift accuracy.

Of course the medium for design—intricate blueprint to hasty sketch, on-site gesture to resolute formula—is symbolic. No doubt the possibilities implicit in the symbols themselves, x-y-z-like, dictate to a significant extent the final design.

Still, the stubborn testimony of facts will confirm it: matter is never strictly rhetorical. Even if facts are a matter of selection, judgment, belief, we beings who live through language often become a bit *too* enamored of our alphabets.

Words, matter, plans—all at the mercy of history.

And if one *can* control a product—of a process, any process—it would seem that the upright species has attained some measure of control over its future, thus securing its thereby disciplined happiness. Provided, that is, said upright species is comfortable inhabiting the reality so designed—walking the walk. A city of bridges, say. Or a nuclear-reactor-dotted landscape. Or a poem lodged, as must all writing be, in the real.

A conjecture: technology as the sum total of these various habitations, hut to skyscraper. To artificial limb.

So I will rebuild, redesign *myself* in order to become the engineer I am not. But to do so requires that, in a fundamental sense, I already *am* that engineer—that I will find within the mettle to embark upon such a profoundly constructive project, to make of my mere self something specifically other. Am I built this way?

Some say you are what you do. But am I what I aspire to do?

Technology as the sum total of these various habitations, hut to skyscraper. To artificial limb.

To self. Design *of* life? An earthly metaphysic? A meta-control? What can you

do. No question mark need apply: shit happens presently, not as it ought to, or might.

20 May 1977. I spot an ad in the *Post-Standard* for a job working as a project engineer in a new brewery north of the city. I apply, get an interview. After six months of traveling for interviews at places like Colgate-Palmolive, in Jersey City—my first airline experience—and Missouri Pacific Railroad, in St. Louis, and Texas Instruments, in Attleboro, this isn't quite the adventure I'd hoped for. Rick is more encouraging.

Just go to the interview and see what's what, Joe.

20 June 1977. Sitting at a steel desk, in a small office, next to another such desk. Atop each desk, a daily calendar and a desk blotter. I share this office with Ted, the only other plant engineer. He started six months before

me. Ted and I share our office with a drafting board, flat files, and a blueprint machine. Ted shows me how to fold and make blueprints of various sizes, size designated by letter—D, E, and so forth. Whenever we use the machine, an ammonia odor fills the room. So we crank open the office windows, which begin six feet up one wall and butt up against the ceiling, affording an excellent view of the generally gray sky.

I adjust my tie—it's choking me. Just a few hours earlier this morning, on my way to work, this same red and blue silk weave—one of two such items belonging to my father—has served as a sign of gainful employ. I glance around at the other drivers, one to a car, most of them men, secure in my sense of place, moving.

At times enthusiastic about the possibilities of engineering, passionate and curious at times about things technical—I enjoy learning about how things work, how to make things work. And I like people, being around people. But I don't understand how to square my passion with what hits me immediately upon starting this job—upon starting every job I've ever worked: the tacit expectation of compliance that permeates human organizations. The chair I'm sitting in, the desk I'm sitting at, the room I'm sitting in, the hours I keep, am expected to keep. And now, sitting in that room, at that desk, in that five-legged, swivel-seat chair—even now the impulse, indeterminate, plays itself out again and again—I flip open my spiral notepad, begin jotting down my thoughts.

I am driven to writing this sense of foreclosure out of my system, plumbing my emotions in the process. Surrounded by the varied techniques of the known, I am by turns literal, lyrical, didactic, desperate to color the sea with crayons. Neither rhyme nor reason in evidence, I, untaught, will put my heart into it, presume to make reason *from* rhyme, words' chance algorithm.

> A time ago
> I found the sun
> and for a while
> I was as one.
> But time refused to let me be . . .

> Refused to let me be.   I do not live here.
> I do not live here.

On my way home, I'm happy to be on the road once again with my anonymous companions in traffic, my tie once again designating my place, my situation, mobile and regular.

While I drive, I try to figure out how much money we owe. I estimate our total debt at around seven grand—rent, light and gas, and loans. My yearly gross is a little over fifteen. I wonder how long it will take me to get us back on our feet, with my father's help. Assuming we stay at 501 Raphael Ave.

I notice a slight shimmy coming from the left side of the '71 Impala's front end. I'll have to get to that.

By the time I walk upstairs, my father is already home. He's sitting on the couch, as ever, smoking, having a beer, TV on.

Hey Joey!—how was your first day?

OK Dad. Not too bad.

What'd they have you do?

Nothing much, really. I walked around the place. Pretty big.

Yeah, I can imagine. Well, so it's your first day!

No big deal.

He's smiling, proud. I'm not sure what to think.

Mike called.

Yeah? How's he doing?

Pretty good. But they've got him working his ass off, poor kid. Lifting these heavy steel things.

Still hardness testing, right? Turbine blades?

Yeah, I think so. He says he might be able to get a certificate at the end of the summer.

Certificate? What do you mean?

I'm not sure—some sorta certificate. You ask him.

I will.

He says he'll be in town this weekend.

Taking the bike?

Not this time. Your mother's coming in with him.

Oh? What did Mom say?

I didn't speak with her, but Mike says she's doing fine. Nice that she could get Mike that job again this summer.

Yeah.

Wish he didn't have to leave town, though.

Well sometimes you've got to.

Yeah. Your mother had to for her job.

Good job, too—you should see where she works. Edison's desk is right across from her, and there's a veranda that overlooks the Mohawk.

Meeng-kya!

Yeah, and the approach to the research center is lined with elms.

Wow. Really something, huh?

Yeah.

That goddamn General Electric though. Burns my ass—

Yeah.

—nearly twenty years and they wanted to cut me down by a buck an hour—

Uh-huh.

—or move to Portsmouth Virginia. Remember? And remember—

Yeah.

—that Jew Hanson, Joe?—

Yeah Dad, I remember.

—remember what that no good fucking Jew bastard tried to do to me?—

Yeah. Had nothing to do with him being a Jew. Can we move it along please, up to the present?

—tried to screw me out of vacation, but I showed him, by Christ!—

Yeah, you were right. Now how 'bout we talk about something that's actually happening?

Shut your mouth! You think you're a smartass now that you're an engineer?

Yeah, I think I'm a smartass.

He's smiling now, and so am I. Lately it's been like this, though. A few minutes about today, a half an hour about yesterday. When he's feeling up, we might talk about the day after tomorrow. But he can handle only one day at a time.

During those first few weeks at work, I continue to scribble down poems in my spiral notepad. I have zero sense of their literary merit, and I'm not even sure why I continue to write. But somehow I must.

My mother has a research scientist friend who's going to Japan, and he agrees to pick up a couple of Nikon F2A cameras at cost for Mike and me. I take mine with me to work, on the way pulling off the road here and there to photograph the weathered trees and barns that dot the open fields along Route 481. I want somehow to capture and communicate a reality more vital than the one I'm experiencing, and somehow I believe my photos will manage this. I even imagine a career in photojournalism.

Doesn't take me long to learn that a camera and film are no less stubborn than a drafting pencil and sketchpad. My photographs are uninspired, awkwardly framed, ill lit.

So I stick to my poems, not having studied poetry a lick. Words seem more pliable, more resilient, even if I have only a young man's eye and ear for them. I'm willing to chalk up my blunders to inexperience.

Still, I can't be sure I'm not a dabbler.

I'm a dabbler.

* * *

During my first month at work, I make mistakes, some of them stupid even to me. It seems I can't help but embarrass myself. I try to learn from my mistakes, but it doesn't come easy.

One morning I walk into the men's room, and as the door shuts behind me, the doorknob falls off—on the inside. I can't work the mechanism, so I'm effectively stuck in there until somebody walks in.

As it happens, on this particular morning, in the large open office area just outside the men's room, the brewery safety office is showing that safety documentary about the guy who wasn't wearing his safety glasses when a grinding disk exploded. The film boasts extended full-color close-ups of the physician picking at the guy's bloody eyes with tweezers and a magnet, trying to locate and remove the metal slivers. Supposed to get you to wear your safety glasses. It's tough to watch.

After ten minutes in the men's room, I end up having to pound on the door and yell for somebody to let me out. Which one of the hourly maintenance guys eventually does, smirk on his face. No doubt one of the men with whom I'll be working on various projects. I can just imagine what he's saying to himself—what I'm saying to *myself*.

Dumbshit engineer.

I have no way of knowing then what I'll find out shortly: that shit will play an even larger role in my on-the-job training.

1 August 1977. Ted and I have been asked to monitor plant water consumption. The wastewater treatment plant, which treats plant effluent before it empties into the Oswego River, has experienced a relatively large increase. And this means extra dollars are going down the drain.

The job consists of setting up a flow meter and flow samplers at strategic effluent points around the brewery. This is what you call *open channel flow*—downhill, by gravity, through large, half-full, underground concrete piping, with lift pumps at the treatment plant to raise the effluent above ground. The effluent contains brewery waste as well as restroom discharge—organics of all kinds. Entry to the underground system is strictly through manholes.

First we pop off the manhole cover. An art in itself, the cover heavy enough to crush your toe if you're not careful. The effluent is generally warm and steaming, vapors rising continually out of the opening. The smell is about what you would expect—the beer from busted bottles in Packaging mixing with brewing spills from other areas of the plant to create a malty, sweet & sour aroma, with a bouquet of shit.

Ted and I take turns climbing the ten or twelve feet down, one hand on the metal rungs of the manhole, the other holding the flow equipment. We're wearing our hard hats, but our safety glasses tend to steam over, so we take them off. The entire inside of the manhole itself is covered with gelatinous scum — buildup from the condensed vapors. If you nudge the surface ever so slightly, you get it all over yourself. Sometimes we wear bibs, sometimes we're in too much of a hurry.

The plant hourly workers — union, International Association of Machinists and Aerospace Workers, or IAMAW, call themselves either mechanics or technicians, depending on their designated "skill level" — these guys walk by occasionally, shaking their heads as our heads emerge from underground. Some of them are close to our age, a few of them I've known since high school, most of them haul in ten grand a year more than we do, with their overtime. Usually they smile, mockingly, shaking their heads.

After Ted and I determine the source of the overflow — the Packaging area — plant personnel locate a water feed whose 4–inch drainage leg, piped directly into the sewer system at ground-level, is wide-open, and close enough to the 120 decibel packaging lines to allow the discharge to go unnoticed. Ted and I continue to monitor flow for four or five months.

Anyone who works in the technical or scientific fields long enough will tell you that technical insight and common sense do not always go hand-in-hand. Whether a response to people or things, our gut reactions, trust them as we may, can be way off. But even common sense will tell you — shit rolls downhill. And if you're lucky, into a clear bag.

As I'll learn, often the hard way: it's all engineering.

During my first year at the brewery, I continue to work closely with Ted, a Penn State chem-E. Ted hails from Harrisburg, has one of the sharpest analytical minds I've ever encountered. We share that crowded office of ours for a year or so before they move us both out, into a large open office area. No windows, no partitions, completely exposed. Individual office construction will come later, two engineers per. And later still there will be talk of returning all workers to an open office space.

Follow the bouncing ball called "progress."

Ted and I work initially for Alan, a former poli-sci major and a good technical thinker who's taking courses nights to pick up his engineering degree. Alan reports to the plant engineering manager, Ron Sparks. Ron is the guy who hired me, a guy with more than twenty years of experience in breweries. Ron is also the guy who, on my first interview, remarks that my hair is a bit long. It isn't, but I get it trimmed anyway in time for my second interview.

Ron understands how the older brewery hierarchy—in which, for instance, the brewmaster *is* master—is gradually being replaced by newer management and organizational philosophies. The influx and evolution of older principles of Taylorized "scientific management" is now coupled with an emerging view of organizations that owes something to newer information technologies. More emphasis, for one, on working interrelationships—between workers and management, workers and machines, salary and hourly, incentive and participation, structure and objective.

Ron is your typical engineer of the sixties—crew cut, spit and polish, decent grasp of technical details, and the customary array of odd personality quirks. He has a certain way of doing things, period. Six months after I first report to work, news leaks that plant management has learned that Ron has falsified his educational records. They bring in another Ron to replace the first.

Christmas week 1977. Just before the Ron scandal, I get a phone call from the brewmaster, Mark Williamson, who explains to me that he has an emergency project he wants me to supervise over the holidays. I've met Mark a few times in the prior six months, and we've had a few uneventful chats. He towers over me—over most employees, in fact. This will be my first job for him, and my only actual process-based project to date. Mark tells me that the wort cooler needs to be modified to meet the new plant production requirements.

Fact: the wort cooler is a large, stainless steel, counterflow plate heat exchanger used to cool freshly brewed, unfermented beer—or wort—prior to adding yeast. Fact: hot wort flows out of the brew kettle into the cooler, on one side of the plates; cold propylene glycol flows on the other side, in the opposite direction. Fact: modification is necessary because the brewery, designed on paper for 1 million barrels per year output and initially built for 3.3 million barrels per, is expanding now to 6.6 million. Projection: at its peak, the brewery will by production standards become the third largest in the world, roughly 10 million barrels of suds trucked through its overhead doors in a single year.

The brewery site itself is a construction madhouse—or utopia, depending how you view things. The local trade unions are happy. And the mechanics, despite their frequent grumbling over the trades doing work they could ("should") be doing, know they have it made. There's plenty of work to go around—for the time being, at least.

Plus, although this is a dry brewery, heavy drinking has been accommodated. In Milwaukee, workers from the dozen odd unions can drink the beer produced in the brewery on their breaks. Lots of accidents and injuries +

lots of insurance money + lots of downtime = a fraught workplace, especially from the company's perspective.

So when they contract our brewery, corporate management sells the union on the idea of a dry brewery, with the stipulation that everybody will get three free cases of beer a month, and up to twenty cases a month at cost. At the time, cost is around thirteen bucks a case. So at the end of each month, you'll see plant workers and clerical staff and management, all with hand trucks, all wheeling cases upon cases of brew out to their cars. I'm no exception. It's a wise policy, in all, and as icing on the cake, the union has also bargained to have the first day of deer season declared a plant holiday.

Alan knows about Mark's request. He's told me in the past to watch myself around Mark. Alan always speaks slowly, calmly, in a deep voice, as if in complete control.

Joe, now listen up: as I told you, Mark has a temper. Just about everybody here's been on the receiving end. Sooner or later it'll happen to you.

Oh really?

Yes. Just watch yourself with him. Just do as he says.

Yeah.

I walk over to Mark's office, on the third floor of the brewhouse, to discuss the project. His secretary, an older, terse woman, greets me with a half smile.

Yes?

Hi, I'm here to see Mark? I'm Joe Amato.

Just one moment.

She pushes an intercom button.

Mark, it's Joe AM-ato.

Seems everybody at the brewery gets my name wrong.

Send him in.

You may walk in.

Thanks!

I walk down the short hall, and into Mark's office. He's seated behind his desk, a huge, hulking man in his late forties with a brush cut, wearing baby-blue company-logo shirt, clip-on tie and navy-blue trousers. The company furnishes all of its salary employees with three pairs of trousers, three shirts, three clip-ons. It's the standard uniform for supervisors who work in the production areas—renders visible the distinction between hourly and salary—and some in upper management, like Mark, make a habit of regularly sporting same to show that they expect as much loyalty from their workers as they demand of themselves. Like most in my engineering cohort, I am no more loyal to the company than I am to my profession, and the corporate duds *do* seem not a little pedestrian, and anything but spit and polish, when mingling with office workers on a daily basis. So I wear them only when baggy pants and true blue seem a good match for the dirty work that occasionally comes my way.

I note as I enter the office that the assistant brewmaster, a nondescript man in his early forties, is standing to the side of Mark's desk, wearing the same uniform. I take my hard hat off as I enter, tuck it under my arm. Mark has a pencil in his right hand, and a stack of instruction manuals on his lap. He's sorting the manuals into two stacks, one in his lap, the other on his desk.

Hi Mark.

Hi Joseph.

More name problems—my given name is Joe, not Joseph, no middle initial. Whenever someone calls me Joseph, I know things aren't going to be easy. But I bite my tongue, and take a seat in the chair directly opposite Mark, on the other side of his desk.

So Mark, what's this project about, exactly?

Here are the instruction manuals for the equipment.

Reaching over his desk and rising slightly, he hands the stack on his lap to me. He's evidently keeping a copy of each manual for himself. Each manual is a half-inch thick. I open a few and look them over—each is chock full of dense technical info and diagrams.

Mark, before we proceed with this project, I just want to be sure that we're designing things the best way.

It's the best way.

But how do you know this? I don't see any drawings.

Mark fidgets a bit, his large frame shifting ever so slightly. The assistant brewmaster looks at me sideways, then does his best to look away. He can see what's coming, I can't.

You don't need drawings. It's a simple equipment addition. I've done this before, many times.

Yes, but, Mark—

The work has got to be done on Christmas Eve.

Christmas Eve? Night?

Yes, that's our only downtime.

Uh-huh. So listen—about those drawings.

I told you—we don't need drawings.

But Mark—I'm the engineer for this project, and I'm responsible for it. And I need to know that this is—

You don't need drawings, Joseph. You have all the information you need right there right in your lap.

Mark, these are simply technical manuals for the equipment you've purchased. This isn't a design—

I'll tell you what the design is.

You'll tell *me*? I'm the engineer, Mark—

You're out of line, Joseph.

But Mark, listen, I'm only asking—

The assistant brewmaster edges ever so slightly toward the office window.

You're out of line, Joseph.

But Mark—

YOU'RE OUT OF FUCKING LINE! I TOLD YOU THIS IS HOW WE'RE DOING IT, NOW GET THE *FUCK* OUT OF MY OFFICE!—

Mark is bellowing at the top of his voice, heaving and seething red, half standing over his desk, volcanic. The assistant brewmaster has nearly turned his back to us, fixing his gaze firmly out the window.

But me—I can think only of my father. Somehow, I'm calm. Mad as hell inside, but calm. I look at him square in the eye, certain.

That's the way you want it, huh?

GET-THE-FUCK-OUT-OF-MY-OFFICE!—

Mark hurls the pencil in his right hand at my head. It whizzes by, missing me by six inches. I'm still looking at him square in the eye.

Fine.

I get up, slowly, pick up my clipboard, and walk out of the office. Slowly.

By the time I get back to my office, Mark has phoned Alan to explain what's happened. Alan calls me into his office, telling me that Mark is sorry, but that my job is to implement the project. No questions asked.

Fine.

Over time I find a way to deal with Mark—past a certain point, you can push him *this* way and he bursts into laughter, or *that* way and he explodes. The working environment at the brewery tends to push people *that* way.

Eventually Mark is replaced as brewmaster, given a technical consultant job at the corporate level. Less stress.

More engineering.

And Alan is soon after replaced by Tom Bilton, a Pitt double-E with a great organizational mind. Tom is an excellent manager with a deep-rooted sense of fairness. He would make an ideal engineering administrator were it not for one weakness: he has little experience with actual design work, so he doesn't really grasp what makes an engineer tick. An engineer, that is, whose passion is technology, and not project management.

Of course, technical insight alone hardly makes for good management skills. And they say that a good manager can manage anybody, any organization. Still, working this closely with engineers means that you would do well to be aware of the technical challenges they're busy confronting day in, day out.

At a more personal level, Tom is another odd sort. He's enlisted in the Marines, I learn, done a tour in 'Nam, become a born-again. He's the sort of guy whose complexion is always rosy, the sort who never curses, the sort who walks around the office area in his stocking feet. The sort whose toughness derives from his cocksureness about people and things. Many of the plant workers don't trust his goody-goody, good-neighbor behavior.

Ted is a born-again too. Both Ted and Tom are married. The two have daily chats about such topics as the decline of the West, the coming apocalypse, and what they term, with due respect for their wives and in step with a cheery, sectarian-friendly AM hit of a year or two prior, "afternoon delight." It's not long before Ted's wife is expecting.

And within the year, Tom ends up an Amway true believer, tries to recruit the entire brewery in the multilevel marketing scheme. This is one of the few times when Tom's personal ambitions compromise his professional aplomb. He holds meetings at his home, a beautiful, vaulted-ceiling structure in Radisson, a planned community with golf course located just outside of Baldwinsville. I attend one packed meeting, decide it's not for me. As I drive home, a cop follows me in my beat-up Impala until I hang a left on Route 31, out of the development.

Working at the brewery is a highly regulated affair. You report to work at such and such a time, you eat lunch at such and such a time, you go home at such and such a time. Which means you piss and shit and shave and shower and eat and sleep—and, as my brother would observe after working as an engineer himself for a dozen years, *dream*—at such and such a time, about such and such things. Taking any profession seriously means tipping one's hat to routine, a routine that will infiltrate your so-called privacy in ways you can't imagine. And in ways you can. When the phone rings at 501 Raphael Ave., instinctively answer with "Engineering—Amato."

And if something comes up—why then you alter *your* schedule accordingly. According to received bureaucratic wisdom—just ask any manager at the brewery, or at your local industrial concern, or at the K-Mart or Wal-Mart down the street—if something comes up, it's *your* job to work it out.

The entire brewery is on a twenty-four-hour-a-day production schedule, seven days a week, three-hundred-sixty-plus days a year. With a couple of days a year usually scheduled for boiler shutdown. And all hourly workers, supervisors and unit managers—many of whom have served a stretch in the armed forces, are accustomed to being pushed around, and tend to reciprocate the gesture—are on twelve-hour shifts, 4 days on, 3 off, 3 on, and 4 off, or 4–3–3–4. Many work 5–2–2–5 and more.

So rest assured, something always comes up.

We're all given beepers, told to carry them with us at all times in the brewery. They beep, and they vibrate. If you're in a high-decibel production area, wearing your earplugs, you can't hear the beep. But you can feel the vibration. Ted jokes that the vibration makes you think you're having a heart attack.

Beeper, telephone, paging system: from heartbeat to heartbeat, they can always locate you.

More engineering.

1 March 1978. The brewing company I work for is at the time engaged in a national PR battle to maintain its place as the second-largest brewery in the nation. Anheuser-Busch is first, always has been. There are some in the company who think we can even surpass the St. Louis giant, because we're now getting ad money from the parent company, a multinational. And in order to be competitive, the brewing company's stock value has to show growth. I learn then that the market analysts whose job it is to assess stock value are called *securities analysts*. One hundred fifty securities analysts are slated to visit our brewery in two months.

Things apparently didn't go too well when the analysts visited the brewery down in Texas. The plant itself was in rough shape. Rumor has it that, shortly after, many of the employees down in Texas reported to work one day to find their personal belongings sitting in cardboard boxes outside of their offices.

Evidently the analysts weren't impressed.

Our brewery—owing to the constant upgrades and related construction—our brewery is a mess. It's decided that having only corporate headquarters personnel on site to prepare the brewery for the coming visit is insufficient. Local plant management wants somebody to take charge of this mess, somebody under their direct control.

And that somebody will be me. I will be the brewery's first buildings and grounds engineer, responsible for supervising all maintenance contracts and implementing capital upgrades to the site. And my first priority will be to prepare the site for the analysts' visit.

Everybody's ass is on the line, including (it is strongly implied) mine. And with the stakes so high, responsibility comes with a certain quotient of power: I will be given virtually complete control over whom I hire, and what gets done.

Doors. Windows. Telephones. Trees. Grass. Sprinklers. Asphalt. Concrete. Rail siding. Paging system. Cleaning. Painting. Roofing. Plumbing. Sweeping and striping parking lots. Signs. Snow removal (though the analysts will visit in late spring). Heating, ventilating, air conditioning. Lighting. Drywall.

My phone rings every few minutes. On my way into work in the morning, I'm trailed by five or six people, each of whom wants something from me.

But I'm having a blast. If I want fifty laborers, I get fifty laborers. If I want to rip out and pave a quarter mile of roadway, I have a contractor rip out and pave a quarter mile of roadway. Without as much as a purchase order, or a "yes" from above. On my authority alone.

It's a frontier, and I'm the sheriff. Or the straw boss, depending how you view things.

I view myself in any case as a good guy. A good guy who, after twenty years of wearing eyeglasses, has just gotten his first pair of contact lenses — hard lenses. I wear them out on the construction site. When the wind blows, I sometimes find myself, out of character, popping a lens out and into my mouth. Takes a while to develop a knack for it, but you can clean and lubricate with your tongue in a few seconds.

One constraint to my marshalling: the company has signed a labor agreement prior to breaking ground on the brewery, agreeing to use only local trade unions to do all of the capital upgrade and outside maintenance. Now the mechanics, like I say, don't always get along with the local trade union workers. But there's another complication, or set of complications.

A large, muscular man approaches my desk. He's wearing a hard hat, dressed like a construction worker. His name is Davy Hills. We've never met, but I know who he is the moment he opens his mouth.

You the engineer in charge of the painting?

Yeah.

I'd like you to take a walk with me. I want to show you something about that painting outfit you're using.

Yeah OK.

His voice is husky, intimidating. I grab my hard hat and safety glasses, walk out of the office area with Hills, down the stairs, out the door, across to the utilities building, and inside. The whir of compressors, pumps, and motors is nearly deafening, but I'm used to it by now. He leads me over to a painter, kneeling down with brush and paint can in front of a CO_2 dryer.

Look at this shit.

He points. The painter turns slightly, catches a glimpse of Hills, but continues painting. You can feel him cowering.

Look at what this asshole is doing. Shit job. Shit work.

Yeah, well, uh—thanks for letting me know.

The Hills brothers, all six of them, control the painters local in Fulton. These guys are real badasses. They put a couple of painters in the hospital early on in the brewery's history for complaining about the exorbitant union dues and the constant harassment (from the Hills) to work at a snail's pace. Since then, nary a murmur of dissent among the contractors, who are forced to use local labor.

The laborers union is another problem.

On any given day, you can find the so-called working steward from the local, Bobby, walking around the site—in his sandals. One look at Bobby, you just *know* that he's not on the level. You just *know* that those dues squeezed out of part-time plow-jockeys are being funneled directly into the pockets of that local's leader, Tom Potts.

Last time I see Potts, he's on TV with his hands behind his back, cuffed, being escorted by two uniformed gentleman through a tall steel gate.

Despite these problems, work moves along quickly, and efficiently. I think well on my feet, know how to hustle, and have gotten down to a science all the paperwork associated with daily work orders and the like. A tour route has been decided upon, and everything looks tip-top.

Two days before the visit, we get a heavy rain. The rain leaks through a small hole in the roof over the stairwell at the end of the tour route. The water makes a mess of the freshly painted stairwell. I bring the roofers in to seal the hole, and find some painters last minute to repaint the stairwell. Everyone is worried it won't dry in time.

I'm on the site at midnight the night before the analysts' visit, checking to make sure all looks well. The paint in the stairwell has dried. On my way out, I run into Ron. He looks nervous.

Hey Joe, how's it all look?

Looks fine, Ron.

Great!

He forces a smile.

Despite a last-minute sewer backup just outside of the cafeteria bathroom (which the expert plumber I've hired, Mack, manages to unplug before any odors enter the dining area itself), the analysts' visit comes off without a hitch. It's so successful, in fact, that Ron receives a memo a week later from the plant manager, Jimmy Plum, commending my efforts in particular, and recommending that I be rewarded with a day off.

My father is happy to hear this, but thinks they should have given me a bonus too.

To: J. Plum
From: J. Amato
Re: Capital Project 007, Security Gate at Wastewater
          Treatment Plant
Date: 1 August 1978

Status of the situation at WWTP is as follows: of the four influent (lift) pumps, pump 2 is operating normally, pumps 1 and 4 are down, and pump 3 is intermittent. I have requested and obtained permission from L. Watts to employ plant electricians to aid the contractor, Ace Electric, in making whatever repairs are necessary to the six (6) 480V electrical feeders and sixty-nine (69) control wires damaged during construction activities. Estimated additional cost is $10,000, but this figure may be a bit low. We anticipate completing the repairs and being back on-line within twenty-four hours. Pumps 2 and 3 are currently handling plant effluent levels, but an increase either in production output or water consumption will overload the pumps. I understand that all production areas have been alerted and are doing what they can to reduce water usage.

The electrical feeders were damaged, as you know, as a result of excavation work that began early this morning. All existing plans of this area, including electrical drawings of influent pump power, indicate that power and control wiring was initially buried at a depth of 48 in., in compliance with the minimum 36 in. corporate specification. As a safety precaution, I had instructed the contractor to excavate to a depth of 24 in. The contractor's equipment cut through the wiring at 18 in. Landscaping activities during the past two years undoubtedly resulted in removal of the (approximately) 30 in. of additional cover. These site modifications were never documented. The wiring could have been avoided entirely only by rerouting the proposed electrical supply. This would have required an additional capital outlay of $5,000, but it was felt both by Engineering and WWTP management that the absence of a security gate system at the wastewater facility was too critical to permit further delays. As project manager, I assume full responsibility, of course, for the final decision to proceed with this work.

I will be sure to keep you informed of all further developments regarding this situation. I am preparing a purchase order and supplemental capital appropriation request to cover the additional funds. These will be submitted to you for your signature as soon as we are advised as to actual repair costs.

JA/jc

Cc: L. Watts
M. DeSanto
R. Smith
T. Bilton

There's something happening here: in this business, a lousy inch (1") can mean your ass, an inch that can result simply from your not having been familiar with a bit of landscaping history. Thus separated from the facts and with only a blueprint to guide you, you end up within an inch of your livelihood.

For most situations, you have to be there, constantly, and at the same time, try not to be the pain-in-the-ass engineer. That excavation?—the contractor used a remote-operated earth saw. Otherwise, we might have had an injury, even a fatality. As the engineer, I should have asked for some test digging, just to be certain.

At least I manage to keep Ted out of it—he had a hunch about the landscaping—and passive construction in the second paragraph helps to suggest that it's not *entirely* my fault, even if it's not anyone else's in particular. Jimmy Plum seems pleased as he reads the memo—with me sitting right there, at 6:00 pm. Jimmy is the sort of guy who walks around the brewery once, and remembers *everyone's* name. He can be firm when he needs to.

Joe, let's not let this happen again, OK?

OK Mr. Plum.

25 September 1978. My brother is working at Xerox in Rochester, on copier fuser development. They're very heavy into technical reports. Having left 501 Raphael Ave. for the first time (aside from summers in Schenectady), he's found a small flat in a quiet neighborhood in the city, a few blocks from the George Eastman House. His landlord lives above him, has the best collection of sixties Marvels I've ever laid eyes on.

Dan and I drive out to visit him once. My brother makes a sauce, dumps in a dime bag of weed. Dan and I forget that he's done this.

Mike, will you pass me the bread?

Mike passes Dan the bread. Dan soaks up the remaining sauce with four slices of bread. In the meantime, I've developed a sudden urge to brush my teeth. I get up, walk into the bathroom.

Ten minutes later, laughing like there's no tomorrow, I emerge from the bathroom with a toothbrush hanging out of the side of my mouth. Mike points at me, laughing so hard he's not laughing. Dan is beginning to glaze over.

We turn on the TV. CHIPS. The three of us take turns laughing. Three-part harmony.

We decide to grab a few beers at the Orange Monkey. Mike drives. Fast.

At the Orange Monkey, we order a round of drafts. Then another. And another. The band is loud.

Hey, get a load of those three babes.

I get up and invite Dan's three babes over to our table.

They walk over a bit anxiously, sit down with us. S-o-p. Only, nobody says a word.

Five minutes later, I look at Dan. He's sitting upright, eyes closed, comatose.

Mike and I have another couple of drafts. Dan's three babes get up and leave.

The next morning, Sunday, I'm seeing triple. I drive the 85 miles back home at 85 mph. I'm still seeing triple.

The next morning, Monday, I'm *still* seeing triple. Starts to wear off around noon.

A new man, Lou, has arrived on the site. A New Jersey contractor has brought him in as a foreman. Lou hires painters from the Utica–Rome local—a no-no, according to the local labor agreement. A no-no, according to the Hills brothers. But for some reason, the brothers don't bother Lou. Nobody bothers Lou.

Lou talks to me as though confiding in me, and I can tell he's confident that my Italian surname gives him some inside leverage with me.

Hey, paisan—koh-ma stah?

Like other contractors, Lou takes Tom and me out to lunch once in a while. On the surface of it, Lou treats us well.

Tom hires a job-shopper techie, Wes, to give me a hand with the swell of maintenance contracts. Wes is a fiftyish, fast-talking, smooth-operating, genuinely friendly guy from Kentucky whose associate of science degree in mechanical technology was sufficient in the early sixties to garner the title *engineer*. Not these days, though, so Wes is having to shop himself around. Lou treats Wes well, too.

But the thing with Wes is that he's hard up for money. He runs on the side what seems to me an oddball business—a machine that he brings around to local shooting ranges to pick the lead out of the ground. The machine gathers lead, along with stone and dirt, and uses a sieve device to filter out the lead, which Wes then sells for a modest profit. I'm concerned that Wes sees in Lou a potential business partner. I've heard him discussing his operation with Lou, who appears interested.

I don't trust Lou, and I try to find a casual way to let Wes know it.

What do you think of Lou, Wes?

Aw, he's a goodoleboy Joe.

Good old boy, huh? You sure about that?—I mean, neither of us knows the guy all that well.

Joe, I've learned to take people at face value.

Just be careful in your dealings with him. I don't want any conflict of interest problems, right?—you work for us, what you do on your own time is your business. But Lou is bidding contracts for us.

OK Joe.

Smitty is a big, friendly, barrel-chested man, always sweating, always nervous. He's my contractor for sweeping and striping access roads and parking lots around the brewery. *My* contractor—I start to think of these men as *my workers*, as employed by *me*, because not only do I deal directly with them on a daily basis, but I also have to take responsibility for everything and anything I have them do.

And because striping is technically painters' work, Smitty pays the Hills a small fee to bring his skeletal non-union crew up from Syracuse. Smitty's equipment is old, but workable. And because Smitty works under contract for me, I have to be aware of this pay-off relationship—I have to know what's going on.

One night, sitting in another contractor's trailer shooting the shit, Smitty begins to talk about his background. He's part Seneca he tells me, has an awful temper, which has gotten him into trouble in the past. He tells me a long, involved story about a guy he knew who, he insists, ended up raping his wife. He confesses to me that he shot and killed the man, and got away with it.

I don't know whether to believe him. But the more he talks, the more he sweats.

Two months after Lou arrives at the brewery, he's standing at my desk, asking me about a new painting job. Smitty appears, and the two men exchange lukewarm greetings. I get up to make a copy of a spec document for Lou, but when I return, Lou is gone. Smitty is visibly shaken.

What's the matter?

That FUCKER, he threatened to kill me!

Calm down, Smitty. What?

I'm not shittin' ya, Joe, he showed me this gun and told me he'd kill me if I didn't get off the site!

Smitty, let's you and me talk with Tom.

Tom calms Smitty down. But neither of us knows what to do.

I tell Wes about it.

You're shittin me Joe!

Unfortunately I'm *not* shittin you, Wes. Watch yourself with that fuckin' guy.

Will do.

I'm never quite certain whether Wes stops seeing Lou outside of work. But I take Wes at face value.

Two weeks later, Smitty appears at my desk. He gives me the high sign, and lifts his shirt. He's wired for sound, the wire taped around his chest. He picks up a pen on my desk, and scribbles on my pad.
fbi.
I nod.
Later, OK?
Yeah.
He pulls his shirt down and walks out.

The next day Smitty arrives in my office, unwired. He explains that the FBI contacted him, out of the blue, about Lou. Smitty told them he'd cooperate and try to get Lou to threaten his life again. But it hasn't worked—Lou's too smart for this.

I tell Tom what's happened. We decide to speak with Stuart Fargrail, chief construction supervisor for the brewery's general contractor, out of Milwaukee.
Stuart's office is in a construction trailer, amid a lot of such trailers that form the south perimeter of the site. Stu is in the Army Reserves, a no-bullshit kind of guy who really knows his shit. When we walk into Stu's office, Tom shuts the door behind us.
Stu, Joe and I have walked over to ask you a simple question.
Tom pauses for just a moment.
What's going on with Lou?
Stu sits back in his chair.
What do you guys *think* is going on?

Two weeks later Tom calls me into his office, tells me that the FBI wants to interview me. I spend a couple of hours alone with two FBI men, each of whom takes copious notes. I'm pissed at Lou—he thinks he can play me for a fool, threaten one of my men while taking me out to lunch. Paisan my ass.
So I decide to tell the Feds just what I think of Lou. I decide to rat him out.

Yeah—a third kind of rat.

I tell the Feds that Lou is a comic-book character.
What do you mean?

Yknow—he looks like he walked right off the set of *The Godfather*.

What I don't tell the Feds is that *they* look like they walked right off the set of a Quinn-Martin Production. What I don't tell the Feds is that this story has been told so many times before, it's now telling itself. They don't need *me* for corroboration—they can simply turn on the TV, or pick up a bestseller. What I don't tell the Feds is that I wish they wouldn't bother me with this fucking bullshit, that they would just get that fucker Lou out of my hair.

The Feds tell me that Lou's best friend, the best man at his wedding in fact, is reputed to be one of the most feared made men in the Northeast.

What I don't tell the Feds is that I know all about made men, and self-made men.

Then the Feds inform me that Lou, who's married, has a girlfriend on the side, Amy—the stunning redhead who does clerical work in the corporate construction trailer, the woman who initially handles all of the sealed bids.

What I don't tell the Feds is that Lou is a bastard when it comes to women. He's told me stories that corroborate this, ugly stories, stories that might help perhaps to indict his moral character in the Feds' eyes. But I figure, there has to be honor among professionals—you've got to draw the line *someplace*.

More engineering.

When I tell my father about Lou, he shakes his head.

Fucking wop guinea bastard greaseball. Stay away from him, Joey, he's no good. Reminds me of a couple of guys my brother Sam once knew. Even Sam wouldn't get involved with 'em.

My father understands intuitively how rubbing elbows with guys like Lou can rub off on you—cause others to see me as just another Italian, just another Sicilian, just another wop guinea bastard greaseball. My father puffs on a Parliament, shaking his head. I try to imagine how he would handle Lou.

I can't.

One sunny day a short while after we let Wes go—Tom has decided that I can handle the bulk of the contracts on my own, and he redistributes a few responsibilities to two new full-time engineers, Vic and Seth—I find myself in the corporate construction trailer, surrounded by a dozen men, most wearing ties but no jackets. At the head of the table is Mikhail Krebnev, the right-hand man of the company's chief corporate construction engineer—

the brewing business's legendary Tari Uralani, a veteran of brewery construction, and a man who is said to know the bulk of brewery design details cold.

Mikhail is wearing a pinstriped suit, and carries with him into meetings the tacit residue of Tari's power. Seated at the other end of the table are two men from Niagara-Mohawk—NiMo, the local power company. One of the men is younger, the other middle-aged, both are wearing suits.

Mikhail is addressing the middle-aged man—who, I learn in the course of conversation, is the chief field engineer for NiMo—rather gruffly. The middle-aged man is articulate, confident, and, well, outright friendly in his responses. He clearly has nothing to hide. There are long gaps of silence between questions. Occasionally one of the other men in attendance, also corporate engineers, poses an authoritatively wrought question. Such questions are greeted with the same combination of expert insight and bonhomie. Me, I've resolved myself to not say a word—far as I'm concerned, I'm the in-house observer.

Turns out that the brewing company is trying to pin the blame on NiMo for worm infestation in the new brewery shrubs and trees, just planted at the not-inconsiderable cost of twenty-five thousand dollars. Their—our—claim is that NiMo's drop, lop & slash technique for installing the power lines through the woods behind the brewery has provided the breeding grounds for the worms.

But this middle-aged man—turns out he holds a master's in electrical engineering *and* a master's in environmental science and forestry. And he's dismantling the allegations of this tableful of corporate engineers with such deftness, and so casually, that it's apparent to everyone in attendance that we don't know what in hell we're talking about, collectively, and that our claim is, in a word, horseshit.

At the conclusion of the meeting, Mikhail instructs me to follow the NiMo chief engineer out into the woods, to investigate the site with him. I don't know a damn thing about worm infestation, but I try to be agreeable.

Yeah, OK Mikhail.

Mikhail and I will have it out on another occasion. But he's OK, and I'll feel sorry for him when corporate operations finally decides to replace him with another suit.

The NiMo man's name is Dave, and he's an entirely jovial guy to be around. It's a beautiful, warm day, with a gentle northerly breeze, and as we walk, the smell from the Nestlé chocolate factory in town mixes with the fermentation odors around the brewery. We're both wearing our hard hats.

As we make our way up the slight grade in back of the brewery, across the rail siding and through a strip of woods to the utility right-of-way, Dave takes off his jacket and loosens his tie. Then he begins poking around through the dead trees and leaves scattered beneath the power lines themselves. He takes his jackknife out, and walking over to a tree on the edge of the right-of-way—a sugar maple, I think—begins to cut into a slight deformation in the bark. Carving a wedge out, he exposes a worm.

This is what's happening, Joe—

And Dave then falls into such an informed discussion of the patterns of worm infestation and tree and foliage growth and the like that, the next day, I'll have forgotten nearly everything.

But one thing I won't forget, and I'll never forget: you don't fuck with the power company, not unless you know what you're talking about. This guy Dave really knows his stuff. And I'm beginning to want the same to be said of me.

December 1978. I've formed my own little construction cadre, a group of workers that are the best around.

Phil the electrician.
Luke the steelworker foreman.
Walt the carpenter.
Jonas the mason.
Bud the pipe fitter boss.
Danny my HVAC man.
Bruce the snow removal and asphalt contractor.

Phil is a stand-up guy, very talented, talks turkey. Ninety-nine percent of the time he's one of the most levelheaded guys you'll ever meet. But he's a bad dude to get on the wrong side of, like most of the construction workers. You learn neverfuckingever to make a bad first impression with any of these guys—you have to earn their respect, and they can be unforgiving as hell. It comes with the work itself—with the terrain, also unforgiving.

So I find that in this world, too, appearances matter. And here the worst appearance possible is a condescending one. If you fuck up in such terms, they'll eat you up alive—in your face *and* behind your back.

Donna works for Syracuse Rigging, their only woman construction supervisor. In fact, outside of a few clerical workers, she's the only woman on the brewery construction site. An attractive woman, in fact, she holds a bachelor's degree in mechanical engineering, like me. I see her in a bar once, after work, and at first I don't recognize her, with her ponytail unwound, her hair brushed out, a touch of makeup.

So she doesn't listen to me, Joe. I'm trying to tell her not to lean up against the panel, because the 480 bus bar is two inches from her hand. But she refuses to listen to me. It's like I'm not even there. Another two inches and she's fried.

So what'd you do?

Nothing. Fuck her, the cunt. She can fry for all I give a fuck.

Phil, Donna's OK. You shoulda told her.

Fuck her, the cunt.

Luke is a great guy. I visit him once, in the log cabin he's built for himself up near Pennellville. He's a soft-spoken, generous man, and his men respect him. When the trunk of my '71 Impala rots out — exposing the exhaust system to the carpeting, and resulting in a fire in my trunk on my way to work one morning — Luke has his steelworkers construct a completely new trunk for me out of sheet metal. Gratis. Lasts as long as I own the car.

Luke subcontracts Walt and Jonas, both of whom are craftsmen, like my father. Unlike my father, Walt and Jonas make fifteen bucks an hour, and benefits. Same goes for Bud, a fitter who can ascertain at a glance every fitting and flange and weld in creation, and who knows how to put together piping systems — to put them together conceptually and manually, palpably — with a sophistication that can only be described as *elegance*. It's a word I'll have a hard time applying to anything I design or make.

Danny is one of the warmest guys on the site. From the moment we meet, I hear in his speech what seems like a slight inflection — I can't say what exactly — and when I converse with him, he tends to lean one side of his head toward me. We get to talking one day.

Where were you born, Danny?

I was born in Ireland, Joe — Dublin.

Dublin! No shit. When'd you come over?

Came here in my teens, in '63.

Just before the war.

Yeah. I was in 'Nam.

What the hell were *you* doin in 'Nam?

Well ysee, alla my friends were being drafted, and I decided I had to go to if I wanted to hold my head up with them. So I became a citizen and went.

You went to 'Nam because your American friends went to 'Nam?

Well, I figgered I had to. I have a sixty-percent hearing loss in my right ear because of that damned mortar fire.

*   *   *

It's with Bruce that I develop the closest relationship. Bruce has taken over his father-in-law's business, turned it into a profitable operation. His two key men are his machine operator–laborer Arnie, a husky man with a high-pitched voice and a crooked nose; and William, a lean, older man, jet-black skin, the lead man when it comes to raking asphalt.

One day we learn that the electrical manhole out in the coalfield, which carries all of the major feeds from the electrical unit substation, has flooded with water. One of the mechanics has walked across what looks like a puddle, and has dropped into a fifteen-foot-deep, water-filled hole with, miraculously, only minor injuries. Also miraculously, he hasn't been electrocuted. We contract Bruce to pump it out, and dig out the sludge.

I walk over with Bruce early one morning to see how it's going. Peering down into the hole, we can hear somebody shoveling, but we can't see a thing.

William—are you down there? Smile, William.

William smiles, and we spot his grin.

I laugh. I know I shouldn't. But I laugh.

Bruce treats his men well. He's non-union (he pays off the local union, like Smitty), but he gives his men a few thousand dollars bonus each year, during the holidays.

I wish to Christ Hanson had paid a bonus. Or James Mahood.

William is amazing. I watch him work—steady, evenly paced. Arnie too. I find out later that Arnie is fifty, and that William is sixty-seven years old. William's body looks to be that of a much younger man. But he has large sad eyes.

Once—only once—my frustration with the turmoil of the brewery, with a paycheck that seems to me lower than what I'm worth, leads me to argue a contract estimate up instead of down. A few thousand up. Thanks to Bruce, three of my friends end up with brand-spanking-new driveways.

Bruce has done the driveways for the brewery executives, too, but they've paid him the small fee he's charged to ease their consciences. Me, I feel like Robin Hood for a couple of weeks. Even here, now, now that I've, what, confessed.

I start receiving gifts at home from a number of my contractors. A Virginia baked ham. Two long strips of filet mignon.

The men sometimes stop by the house when I'm not home. Steve, from the cleaning company. Bob, my insulation man. My father invites them upstairs, has a beer with them. The next time I see them, always the same question.

Hey Joe, I spoke with your father the other day. Nice guy. Tell me—how long you been living there?

A decade or so.

Huh.

Two months later, corporate management puts out a memo forbidding employees from receiving any and all gifts from contracting firms.

And it's decided, after another few months, that the contract situation at the brewery has become too much responsibility for a twenty-three-year-old.

Good thing. Even setting the Lou debacle and others like it aside, I'm being harassed daily for everything under the sun. They're running me ragged, from resetting a circuit breaker because a secretary has plugged in a space heater, to carrying a bag of salt over to the guardhouse because the sidewalk is icy. I often end up doing these things myself simply because it's just plain easier, and quicker, than handling the paperwork.

I've become the brewery's unofficial ombudsman.

And besides, the plant is growing up, and clamping down on spending. Almost overnight I find that an approved purchase order is required prior to my awarding even a hundred-dollar contract job. That all jobs over a thousand dollars must be handled as sealed bids. That my expense budget will now be subject to another layer of management approval, and that the Return-On-Investment on capital projects—new parking lots, new offices, and the like, more and more of which are being planned a year or two in advance—will be lowered from five years to three years. Et cetera. The once and future corporate way: internal fiscal accountability becomes *the* priority.

End of frontier.

It was never a frontier, of course.

And another thing.

I'm walking along the backside of the brewery at the end of a long summer day, and a construction pickup passes me, slowly, on the perimeter access road. The truck is filled with workers on their way home, sitting in rows on each side of the bed, hanging onto the overhead equipment racks. As the truck passes, a familiar voice yells out to me.

Hey Joe!

I look up, but can't pick the face out of the crowd of hard hats.

Hey—it's Rich Delmedico Joe!

Now I spot him—it's Mr. Delmedico, Rick's father. He waves as the truck continues down the road. Instinctively I wave back, hesitating for a split-second before I call out to him, caught like a bug in urethane—

Notes toward a Supreme Fiction

* * *

Because I feel awkward as hell. Because on *this* site, I call my men by their first names, and I insist they call me by mine—this is the only way to work with these guys.

But Mr. Delmedico will never be Richard to me, or Rich.

So here I am, on the job, with an opportunity to puff myself up but good, address this man I've known since I was a kid by his given name, a man who knows more about electrical work—about *work*—than I know about anything. I mean, what do I know, really, about a lifetime of work? What do I know save for moving a few men and machines around a job site, and filling out the appropriate forms? Mr. Delmedico?—a man with a shitload of experience, day in and day out, earning a living for his family. Me?—with the exception of a few design-based projects like that WWTP security gate, which I manage to fuck up but good, I make clever, shoot-from-the-hip decisions, yes, I manage to coordinate people and things, OK. A talent, and things do get built. But I don't even call the big shots, which are left to the big shots to call.

So the opportunity to address Mr. Delmedico by his given name comes *not* from my having attained full-fledged adulthood—I'm not even sure what this might mean—and *not* from having proved myself particularly knowledgeable or capable. Hell no. This opportunity is derived from what seems to me, in this conflicted instant, little more than my corporate, straw boss status—

Hi Ri—Mr. Del—medico!

I smart from this little incident for weeks, fragile lad that I am. Am I what I aspire to do, to be? Whatever. Glued in place, I could not be what I had not yet become.

What I would never fully become, if becomings are measured in lifetimes.

It's around this time that Rick, Frank, Stan, Mike, and I and the rest of the crew from Dolores Terrace have starting hanging out occasionally at Orpheus, a disco over on Butternut Street on the north side, a place that used to be called The Shack. Disco has hit the Salt City big-time, and though most of us like some of the music, most of us are not sure what to make of the glitz.

At Orpheus, though, we can dress the way we like—jeans, T-shirts, whatever. José the bartender just keeps pouring those Tequila sunrises and jellybeans, and we just keep coming back for more. Rick delivers parts for the large Chevy dealership in the village, and his friend and coworker, Slim, hangs out at Orpheus. Slim is a bit older than us, pushing thirty. It's through

Slim, in fact, that Rick learns of the place. To us, Orpheus is a low-stress environment—we don't have to be on our guard, doubling our fists and flexing our biceps at every raised voice.

Only one thing: Slim is bi, and seems to be struggling with his sexuality. As a matter of fact, a number of the regulars at Orpheus are either gay or bi. You might in fact call Orpheus a gay bar, which at this time and in this place means *overflowing with polyester and pose*. Calvin, the DJ, dresses on occasion in pink negligee. Tammy, the waitress, is in reality Timmy (the give-away comes when you're standing beside him at a urinal). And many of the women who hang out at Orpheus tend to dance only with women. In fact, Slim hits on all of us, usually ends up the evening completely wasted, drunk on his ass. Sometimes we do too.

Remember—this ain't the City by the Bay. We're not quite sure what to make of all this. So we keep our anxieties to ourselves for the most part, and we take turns flexing those biceps, oblivious to what beefcake signifies in these environs.

But for a bunch of aspiring ladies' men who spend way too much time idolizing Harry Callahan, who are far too familiar with Evel Knievel's latest exploits, well: we poor macho slobs find ourselves for a spell curiously entwined in gay life.

Timmy, move the fuck out of the way.

We're moving Bobby, Timmy's sometime lover. Not much stuff really—two dozen boxes and some light furniture. Timmy is wringing his hands, getting in the way. Bobby is not much help either.

Slim, you're such a beast!

Mike, Rick, Frank, and I are having a hard time keeping a straight face.

We get in our car to follow Slim. Slim is driving the Wells Fargo van, Timmy in the passenger seat, Bobby in the back to keep things from moving around. We're up on the north side, heading down Lodi.

Suddenly Rick spots a fire under the van. Then smoke. The four of us are waving and yelling *Pull over! Fuckin' pull over!* out the car windows, trying to get Slim's attention. We're also laughing our asses off. When Slim finally notices, he swerves the van toward the right side of the road. Timmy leaps out of the van while it's still rolling, and ends up face-first in a telephone pole.

Slim jumps out of the van, Bobby bursts out the back doors. But the fire puts itself out. Frank kneels down to have a look under.

Busted fuel line.

Timmy seats himself on a sidewalk step, his bloodied nose cupped in his hands. Bobby is hysterical. Slim is good and pissed.

Bobby you fucking faggot—shut the fuck up!

The rest of us are still laughing, trying not to laugh. An old woman appears at her doorstep. She surveys the scene, shakes her head, walks back inside.

Mike, Frank, and Timmy begin to unload the truck on the street, Rick and Slim walk off looking for a pay phone. I walk over to Bobby, offer him my hanky, chuckling.

He looks up at me, takes my hanky, nods toward the van.

Thanks. They don't build 'em like they used to, huh?

His voice, his mannerisms have changed completely—suddenly he's just a guy like me, no frills. No matter how many times I hear Bobby switch like this, I'm startled.

No, they don't. Corporate America. Who came up with that term, anyway?

Fucked if I know. Goddamn corporations jerk-off the unions, and the unions jerk us off.

Yeah, well—we need unions, right?

Damn straight, I'm no Republican. But we need good vans too.

Yeah.

Things taper off with Slim and the rest when it becomes apparent to us that hanging out at Orpheus with gay men is not exactly increasing our chances of getting laid. For me, this comes as something of a relief—the engineer in me, like the straight guy in me, seeks only clear-cut, neat and tidy solutions to life's little problems. Like good and bad desires, there are good and bad designs, and good and bad choices. This is the reality of engineering life, the engineered real—which is, as a rule, only as real as compliance.

So if there's a queerer reckoning in the cards, it'll have to wait till I lose the calculator.

By the holidays, our debt at 501 Raphael Ave. is completely paid off. It took a year and a half of brewery paychecks. We order cable TV, just in time for Christmas. Our two-foot-high artificial tree sits atop the TV set, a single string of small lights and a dozen tiny bulbs twinkling while my father sleeps.

31 May 1979. We've got new office digs, and I'm sharing my office with Chris. I've been reassigned, given the responsibility of all utilities system upgrades—steam, air, carbon dioxide, water, ammonia refrigeration. I'm finally doing design. Or redesign—reworking what's there to make it better, somehow.

A bit like refinishing. Was this lost on me at the time?

Chris is in his late fifties—my father's age. He's worked for Allied Chemical in Solvay for twenty-five years, is married, has two sons and a daughter a few years older than me. His daughter will die of cancer twenty years later, as Chris and his wife approach eighty.

Chris earned his engineering degree, chemical engineering, at Syracuse University, on the GI Bill. Was at the Battle of the Bulge. Infantry. Stocky guy, getting fat. One eye a bit damaged—guy hit him with a beer bottle during the war.

The first six months I know him, I think Chris is a real fucking asshole. Whenever we exchange a few words, he won't say much—just squint, and laugh. Cryptic as a motherfucker.

Why'd you leave Allied?

Well, let's see. They took the biggest gun they had, pointed it at my head, and told me to leave.

Chris smiles, squinting, almost winking at me.

But after a few months, I begin to realize that Chris is just a bit shy. And that he has a wicked sense of humor. And that he knows, he really, *really* knows his technical shit.

But he's uncomfortable meeting people, completely uncomfortable with managing people. When Tom takes vacation, Chris refuses to serve as temporary engineering manager. He's resolute—he just won't do it. And Chris is much older than the rest of us—there are eight of us now, and three draftsmen. So Tom is forced to pick a younger man.

Chris has spent a lifetime designing the many processing operations that have to do with sodium carbonate production at Allied—the Solvay Process. He understands both the theoretical and the practical aspects of control systems, whether pneumatic, electronic, or, more and more the case these days, digital. How and when to open valve w to admit x amount of substance into vat y through piping z at temperature a. He understands, inside out, the substances, the reactions, the equipment, and the underlying concepts, *wxyza*, and then some. He's taken a number of grad courses in chemistry, holds close to a master's in that science. There's nary a technical question I ask him for which he doesn't have the answer. And on those rare occasions when he doesn't have the answer, he knows how to work one out. And he makes *me* work for it, he *always* makes me work for it—use my *own* head.

Chris, wouldn't the flow go up if the pressure at this point goes down? C'mon Joe.

He smiles, squinting. This is Chris's way of telling me to *get real*.

One thing Chris doesn't know the answer to: why they say they eat cats in Solvay.

Has something to do with the poverty, I think—before the war.

After a while it's clear to everyone that Chris is teaching me the tricks of the trade. As in the trades, he's the master and I'm the apprentice, struggling to be a journeyman. One of the draftsmen, Travis, sketches a picture of me sitting at my desk, with a bookcase hanging on the wall above. On the bookcase are several engineering manuals, and a nice caricature of Chris, sitting on the bookcase with his hands folded, looking pleased as punch.

This is around the time of the Three Mile Island mishap. Ted, generally unruffled, exhibits some real concern. His family lives just outside of Harrisburg. The rest of us compete with one another for the most insightful event projections. True to the spirit of the region, we wax apocalyptic. True to our profession, we exude steely composure in our most dire proclamations, quibble technically over the extent to which the reactor core has been *compromised*. In truth, we're all a bit enthralled throughout the entire episode. We just can't help ourselves.

I join the Instrument Society of America. I decide that I will make instrumentation and control my subspecialty. If I've got to be an engineer, I may as well make the most of it, fuck it.

Turns out, unsurprisingly, that Chris, the ISA secretary, is a local legend among the instrument and control people in Syracuse and around Central New York. There are a number of men—seems it's always men, and white men—who have become known authorities among the technical sales reps and the engineers in the various manufacturing plants. Chris has a rep among those who have made the rounds.

I begin to dig deeply into instrumentation and control system design and redesign. Temperatures, levels, flows. Ball valves, gate valves. Stainless steel butt weld piping, threaded iron piping, flange fittings. Steel tanks, plastic resin tanks. Centrifugal pumps, positive displacement pumps, peristaltic pumps. Electric motors, motor starters. Power distribution wiring, control wiring, conduit. Circular recorders, strip chart recorders. Gauges, relays, switches. Phase changes, pressure drops. Mixing, metering, calibration. Condensate, steam traps. Orifice plates, turbine meters, vortex meters. Solvents, acids. Gases, superheated vapors, saturated liquids. Continuous processes, batch processes. Proportional, integral, derivative control.

Take a simple globe valve with pneumatic actuator, used to supply coolant to prevent the temperature of an exothermic process—such as fermentation—from rising above a given value. If the temperature of the process rises a specified amount above the established set point, the valve opens to provide more coolant. This obviously requires some sort of temperature measurement device that sends a signal to the controller—the mechanism controlling the valve. Simple enough, on the surface of it.

But controller output controls, of course, valve operation. And accurate and repeatable (*repeatable* means, for us, precise) control of temperature generally requires something more sophisticated than off/on, open/close control. Clearly, the higher the temperature (and corresponding measurement), the wider and/or more frequently the valve should open. We call this *direct*, as opposed to *reverse*, action. When the temperature measurement drifts above or below set point, we call the result *error*. So three standard questions, relating to three standard controller output modes: (A) should the output of the controller change percentage x in proportion to the percentage change in error? Or (B) should the output of the controller change percentage x and y frequently in proportion to the percentage change in error? Or (C) should the output of controller change percentage z in proportion to the time rate of change in error?

We call A, B, and C, respectively, *proportional*, *integral* (or reset), and *derivative* control action; and as one might have guessed, there are mathematical functions corresponding to all three.

In time I learn to ask questions that are as practical as they are theoretical. What happens with various combinations of A, B, and C? When should you use mode C, for example? When shouldn't you? Given mode A, if we call *gain* the relationship between percentage x and percentage change in measurement, what should the gain be? And how, exactly, do you adjust gain? And what do we mean by damping?

Whether A, B, or C or x, y, or z—not so simple, on the surface of it. This is what you call Chapter 1, the basics. And in seven years of industrial experience, I never get to explore much more than the basics. So I only momentarily belong to that "we," above, and it only momentarily belongs to me.

We, they, I—shifty pronouns, shifting together, sifting life's mixed loyalties. I'm never sure what to make of *them*. And even the most technical discourse registers their presence or absence, any claim to accuracy complicated by the variegated truths of such symbolic things.

More engineering.

My first major project has to do with upgrading the instrumentation and control system used to fire up one of our large natural gas boilers. A bit risky, because boilers have been known to blow if everything isn't sequenced correctly, and if the proper safety apparatus is not in place.

A few months earlier, another boiler had in fact exploded. Or to put it another way, the boiler operators blew it up. I talk with the unit manager, Ray, ask him what happened. Ray served in the Merchant Marines for a decade. He shakes his head, explaining between puffs on his pipe how lucky we are that nobody was killed.

Why, that boiler looked like a dog bent over shittin' razor blades.

My boiler redesign works out OK, thanks in large part to the plant instrument and electrical staff—the hourly union technicians who are actually doing the work, and their managers. Many of these guys are sharp as hell. We get along well, but it's a bust-the-engineer's-ass arrangement—in my particular case, good-hearted. When they pass through the boiler room during our initial attempts to bring the boiler online, many of them break into a run, covering their heads, tool belts flip-flopping.

Eventually I take over a project from a coworker who's landed another job—I'm to use his half-finished drawings to redesign the carbon dioxide reclaim system. Carbon dioxide, like alcohol, is one of the by-products of fermenting wort to produce beer. Breweries attempt to reclaim as much as possible in order to reduce the amount they have to purchase for carbonating beer further along in the process. Chris holds my hand at first, as do the hourly workers. Once in a while they'll catch me with a tool in hand, doing a bit of manual work myself.

Man, you're just *achin'* for a grievance, you cocksucker you.

They leave me alone, though. Which isn't the case when I walk into the production areas. The workers in other departments eye me carefully if I as much as open a valve to test for flow-through. So I learn where to be ambitious, and where not.

After a while, I know pretty much what I'm doing. I understand the better choices to be made among the finite possibilities before me. I understand, that is, engineering *as* an engineer. Which is to say, as a technical professional *and* as an employee of a corporation.

Only one problem: I'm handling anywhere from two dozen to three dozen projects at any given time. Back in Chris's day, the early fifties, when he was learning to become an engineer, he might have worked on, at most, a half a dozen projects at a time. From a strictly technical point of view, he really had a chance to learn what he was doing, the details. Whenever I ask him about his younger years, he shakes his head.

I was always told to try to make it a little better than the guy before me. These days they want managers, Joe, not engineers.

But we have calculators and computer programs to help us, Chris. You guys worked with slide rules.

Chris clears his throat.

Joe—when something goes wrong, you have to know why. It doesn't matter then what tool you've used to design it—you have to understand the principles behind the processes themselves, and the way the equipment works. Good engineering requires this deeper understanding.

* * *

At the time, I give a lot of thought to my math degree. I wonder what value it has in my work, a good portion of which is, in effect, not applied physics, but applied *engineering*. I wonder whether my background in math contributes directly to the "deeper understanding" Chris counsels. In my entire engineering career, I will use differential calculus only once—to take a first derivative in order to solve one of the monthly quiz questions in *Power Transmission Design*. Gains me some respect among the other engineers in my department, but that's about it.

Still, all said and done, it's not the facts or methods, exactly, that I seem to have learned in my academic studies. It's more a certain persistence of thought, and an ability to think in ways that open the familiar to another layer of insight. To identify patterns—relationships among objects, networks of relationships. Perhaps to create or assemble such relationships—an endeavor as imaginative as it is rigorous. And occasionally, to troubleshoot—to spot the singularly unfamiliar in the familiar, to locate the detail that doesn't belong.

That is, if I have the time.

Even so, without my native curiosity, I'm not sure *where* my more analytical side might have gotten me.

Or my more emotional side. Funny thing: two of the technical areas in which I develop a keen interest, areas particularly pertinent to instrument and control work, are pumping and piping, and thermodynamics. Now, I managed only a grade of C in both of my thermo classes in college. But somehow my inability there has translated to asking the right questions on the job. I'm prone to probing further and further into the sticky interrelations inhering among phase changes, heat transfer, energy levels, and the like.

As to pumping: the only hands-on experience I have with pumps is removing & replacing automotive water pumps. Otherwise, I have a rather limited experience of my *own* circulatory system. Fact is I prefer not to think about pulses—I can't even *read* about blood flow without wincing—and I don't like touching the insides of my wrists. A phobia, I guess. It's not that I can't stand the sight of blood—the sight of blood I can tolerate, to a reasonable degree. But.

An admission, another confession: I passed out once taking my pulse, in my high school biology class. One moment I was feeling for a regular beat, trying to establish a count. Next thing I knew my ears were ringing, my head resting against the cool vinyl asbestos tile.

Somehow though, when it comes to piping systems, I'm entirely drawn to them. Go figure.

Most of the industrial systems I work with utilize centrifugal pumps, which essentially impart velocity to a given liquid. One needs to know how to calculate pump output according to liquid viscosity, and according to the system in which a pump is installed. Yet another pronoun—*one*. Even now, I can recall that a centrifugal pump operates at the intersection of the pump curve and the system curve, each curve a plot of head pressure versus flow.

Gotta talk to that boy Danny. He's eatin' his own fuckin'.
Yeah. At least we got him comin' down on his wife. But Jesus.
Despite the salary–hourly distinction, I'm often privy to the instrument and control technicians' daily discourse. These guys can be fastidious as hell when it comes to installation details, but the ebb and flow of their shoptalk leaves one marveling at the workaday habit of speaking one's mind. About just about *anything*.
You guys ready to talk about the job?
Sure. Hey Joe—you ever eat your own fuckin'?

As part of my professional training, I'm sent to a two-week instrumentation and control training seminar in Foxboro, Massachusetts, all expenses paid. I'll be going with two other guys in the plant engineering department, Ray and Jerry. Jerry is an instrument and control unit manager—a chubby guy, computer science grad back in the late sixties, then drafted and sent to 'Nam to work on military information systems. Turned out one of the chopper crews needed a guy on machine gun more desperately than a programmer, so they recruited Jerry. He flew only eight or nine missions, though, spent most of his time sitting on his ass. Waiting.

Whenever I ask Jerry about his time in 'Nam, a nervous giggle accompanies his narration. Having found himself a bit displaced from his preconceived duties, he nonetheless developed an appreciation for the machinery.

Fucked if I know how I ended up there. But that minigun was some crazy shit, man—a coupla bursts'd level jungle, hooches, gooks and all.

At Foxboro, the three of us spend a good deal of time with one another. These two guys are sharp, and a blast to be around. On the first day of the seminar, we decide to eat lunch at a nearby Friendly's Ice Cream. Ray orders a burger and fries.

How would you like your hamburger cooked, sir?
Just take the pulse out of it.
Jerry smiles, leans over, and whispers in my ear with that nervous giggle of his.
Just wipe its ass and send it on in.

Later that night, in the hotel bar, Ray and Jerry are both on the make. Like most married men, they're trying too hard, though you can never quite tell if they're serious. Me, I've got a girlfriend, Cindy — nice Italian girl from a large Italian family — and I'm not one for screwing around. Still, I strike up a conversation with a pretty woman, around my age.

So what do they need engineers for in a brewery?

I'm used to being asked this question, but I'm never quite sure how to handle it. I know I can't talk technical details with most folks. And even with those who have a good mind for such details, things can get boring fast. So I tend to talk about the *things* I work with and around — the equipment, the materials, the stuff that makes what I do *feel* the way it does.

Well, there are cereal cookers, and mash mixers, and brew kettles, and what they call lauter tuns, and all sortsa shit that—

Well — I guess then you help to make better beer and shit?

Uh, yeah — yeah, I guess so. And maybe make it more cheaply.

Uh-huh.

And maybe more safely—

Yeah, I get it — your job is all about beer and shit. You're a beer and shit engineer!

She laughs as she says this. At that moment I feel very much pinned and wriggling on the wall. I'm not the man they think I am at all . . . .

Back home, I attend an instrumentation show at the Marriott over on Carrier Circle. It's a crowded affair. Mulling around, I spot a familiar crop of red hair. It's filthy-rich Kent — my friend from my undergrad days. I haven't seen him in a couple of years, didn't even know he was still in the area.

Hey! How are you man?

Hi Joe.

We shake hands. He's smiling a bit weird, distant, something seems off to me. We walk and chat a bit, pausing in front of a booth every now and then to examine a new piece of hardware.

So how *are* you Kent?

Doing fine.

Yeah?

Yeah. I've found Jesus.

I ask for his business card and give him mine, just for the hell of it.

August 1979. I'm not the only one who makes mistakes.

The mechanics on one occasion double as plumbers, fitting a tank of 150-lb. compressed air to the first-floor plumbing system to clear a plugged drain on the office building second floor. This is what you call *not thinking*.

Without knowing it—for the ten minutes or so it takes before someone has to use the bathroom—water, toilet paper, tampons, and shit comes shooting out of the sinks and toilets all over the second-floor bathrooms. Lucky no one is sitting in a stall at the time.

On one occasion, two mechanics even use an *oxygen* tank to clear the lawn sprinkler lines. In an oxygen-rich atmosphere, the world is dangerously combustible. As Chris later observes, lucky they didn't blow up the brewery.

But there are more subtle matters at work here.

For one, to be a mechanic *or* an engineer is to inhabit a world divided, or thought to be divided, into two discrete domains of experience: *texts* and *things*. Engineers are hardly literary types, but they customarily see the world more in terms of *text*; mechanics more in terms of *things*; technicians would appear in some sense to straddle this divide; whereas managers would seem to make of these texts *and* things a matter of methods that will presumably relate, somehow—or be made to relate—to the organizational human. For managers, texts and things are transcended by organizational motive—by business imperatives. For engineers and mechanics and technicians, the motive is *in* texts and things. For all of these workers, the social contribution, good or bad or in between, is often lost in the shuffle.

Working with text requires an ability to tease out patterns—as in comparing that last sentence with this one, in which instead of resorting to the former omniscient "voice," I hereby impose the pronoun *I*, the customary mark of self; whereas working with things requires an ability to work with tangible bodies, in three-dimensional space—as when you find this book an item of minor heft.

When I consult a process and instrumentation drawing, or P & ID, I must visualize the hardware, the equipment, as interrelated *things*, represented on such drawings as, and through, often elaborate symbols. And accordingly, I can in general make two types of mistakes in laying out my design: I can communicate inaccurate information, or I can communicate inaccurately.

In order to communicate accurate information, I need to understand the thing itself *as* a thing. I need to be aware, for example, that the brewery's Cold Service area—which includes Fermentation, Aging, and the corridors where wort and beer are stored and filtered and carbonated—is kept at approximately 36 degrees F. And that the underbelly of various kettles and the steam-related equipment that forms the Brewhouse first floor, just on the other side of the heavy refrigerator doors that help to keep Cold Service cold, is often over 100 degrees F. Plant workers who must work in both areas are permitted a twenty- minute adjustment break. To fully grasp these working environments, I must myself be a part of them—I need to spend some time there, because these extremes of temperature and humidity will affect not only the materials required by a design, but *how* the work will be performed in order to carry out a given design. I must know what I can expect from workers working under these conditions. I must know what workers can do to ensure quality construction, and in some cases—as in welding around explosive grain dust—what they can do to protect themselves and others from serious injury.

At the same time, in order to communicate accurately and effectively, I need to understand language itself *as* language. Not simply the shared symbols and writing conventions of a technical community, but the vagaries of meaning, of symbolic variance and variation, with which these oddly shaped scrawls enter into our thoughts, and help to shape thought itself. There are innumerable ways of presenting a given design, or a given set of instructions. And no set of standards will automatically do the work of addressing specifics of context and audience, of layout, of negotiation with others. I need to be aware, for example, that Zack has been promoted to a supervisor from the hourly ranks. And that many of his former union coworkers regard him as a turncoat, and pick on every minor detail of a project for which he is responsible. I need to anticipate zero generosity here in order to be a bit more descriptive than I might ordinarily be in my write-ups and instructions.

I'll have Travis and the other draftsmen to help me with my designs, and to visualize them fully. I'll have our two secretaries, Joan and Jan, to proof my writing, and to suggest revisions. I'll have Tom looking over my shoulder, and Chris and the other engineers to give me a hand, and plant workers with whom to collaborate on final design while they do the installation. But *if* I'm to take my job seriously, *if* what I'm about is something more than bookkeeping, and spreading blame—I must assume final responsibility for this sequence of cause and event. Somebody must, everybody must, I must.

To be sitting at one's desk with a blueprint, thinking of processes and products, of how things work, together; to be out in the field, with things, perusing and exploring, with a mind to seeking out the best representation that might embody a new dispersion of things into the world, a dispersion that itself enacts actual shifts in density, phase, viscosity, flow—these activities require, as much as anything else, a concerted effort of imagination. Theories of text, practice with things; theories of things, practice with text—theory and practice both, what I'll later understand in more scholastic confines as *praxis*. There is training, of course—on the job. Yet the sort of education required here is anything but rote. Without imagination, without curiosity, without the persistent application of inquiring thought to people and things, often with the aid of text—without these, there's not a chance you'll be truly good at this kind of work, *your* work. No way.

And even if you are truly good at it—in a world of control grown so remote and so sophisticated, on a planet of social collectives grown so interconnected and at odds—it's far from self-evident whether such work does any real *good*.

Don't I know it. Damned if I don't know it.

December 1979. I'm working long hours, lots of projects. I'll drive home in the dark, after a long goddamn day, completely exhausted. My father will say the same thing each time.

You look tuckered out, Joey.

Sometimes I'll collapse in bed without eating dinner. My father will wake me up the next morning after a deep, ten-hour sleep. Eventually, after a few more days of this, he'll wake me up and I'll be congested, head aching, feverish. And neither bed rest nor my father's cooking will lick it. After three days of bed rest, I'll end up at the doctor's office.

The doctor will tell me to take it easy, again, and again prescribe antibiotics.

Take it easy, yeah, OK. Craftsman Interiors isn't doing too well, and my father's full-time job is looking shaky. More and more it looks as if I'll become the breadwinner, head of household—which is the last thing I want. I want, in a word, *out*. Out of this profession, out of this life, into something more—

I can't say, exactly. I have no idea what my options are, however finite. Even the joy of sheer knowing, of learning something I didn't know, now seems somehow not enough.

I try to work myself into better shape, working out more often, jogging. I pull a back muscle.

Chronic fatigue? A decade later, when the same shit starts happening to my brother while working at GE, the medical community will land upon a similar expression to describe a rarer, more serious disorder: Chronic Fatigue Syndrome.

Two words or three, the future begins to appear a matter more of probabilities than possibilities.

The IAMAW threatens to go on strike. We've now got five engineers who are in their twenties—Ted, Vic, Seth, George, and myself—and my four peers all regard the union position with contempt, as sheer foolishness. Chris has mixed feelings, having witnessed firsthand a number of strikes, having seen the damage they can do.

Me, I'm not sure what to think, though I think the union *is* pushing it some. We all understand that this threat is fueled by dissatisfaction in the production and warehouse areas. The plant engineering hourly personnel—instrument and electrical technicians, and utilities mechanics—these guys oppose the strike, but are ultimately out-voted.

So for six weeks, salary workers congregate in a Dunkin Donuts parking lot at 6:30 am. The plan is to work twelve-hour, five-day-a-week shifts to keep the brewery running. Management figures it's best to show up at the plant in force, so it's donuts and coffee together for courage, and then we're all to drive to the brewery a mile down the road.

The Dunkin Donuts lot is jammed, so I pull up along the curb, idling. Tom sees me, walks over with a chocolate glazed in hand. I lower my window, he briefs me on the details. Our every breath is visible. *Cold as a witch's tit*, I say. *You bet*, he says, smiling, upbeat as always, oblivious.

As we begin to make our way down Route 57, I feel a bit like I'm part of a funeral procession. I turn into the brewery, driving my car slowly (as instructed) through the picket lines, the strikers yelling "scab" and beating on my hood. A few of the guys shout "management fuck."

It hurts, being called a scab—the very word my father used with such disgust back on Dolores Terrace in the early sixties, when his union would walk out every few years. Technically speaking, I'm not a scab—but it hurts. And as to being a management fuck, well—I'm not any more management than are they. Fuckin' pisses me off, as I say to Chris, because my sympathies are with the workers, in so many ways. He nods.

Anyway. Plant management feeds us well, pays us all sorts of overtime for our trouble, even creates a bonus plan to augment our base salary during the strike. So it's a battle over loyalties, as ever, even if the lines are drawn in the sand.

In the first week of the strike, on my sixth twelve-hour day in a row, I pull my Achilles tendon running up a stairwell in my work boots.

And that's it for me. The doctor tells me to stay off of it. But I limp around the plant, feeling like a—well, like a stupid asshole. It'll be months before it's healed enough to do any manual labor.

Me, of all people—I ran myself ragged, forgot to pace myself.

Two weeks after the tendon pull, I'm diagnosed with bronchitis for the third time in six months. Now I have no choice but to call it a day.

The strike ends after six weeks with the union capitulating. This doesn't set entirely well with me, given my own convictions about collective bargaining. And my father's. Of course my father is a strong union loyalist, as I say. But even my father has bitched about unions—the problem with seniority, and with wanting more than you're entitled to. Unions are like companies themselves, and inherently imperfect in this regard. So I end up reconciling myself to the fact that the strike was a stupid move, strategically speaking, motivated more by the bitterness of dead-end jobs than by abusive company policies.

And during these months, it seems I'm forever being sweet-talked by supervisors to make the move out of engineering design and into management. Despite, or maybe because of, a number of run-ins with coworkers and plant personnel over customary factory issues—confrontations such as the one with Mark Williamson, general one-upmanship, daily badgering, my refusal to suck-up as a company loyalist—I've developed a reputation as friendly, down to earth, and outspoken. In company lingo, strong leadership qualities. In more colloquial terms, I'm no asskisser. The company seems to respect this attribute, even as it wants to mold me into the very thing I strenuously resist—a company man.

But I'm adamant. I want to stick with design, I say, supervise only the installation of my own projects. And besides—I've had my fill of invisible leashes, wiping everyone's behind. While in charge of buildings and grounds maintenance, I'd been on call day and night and holidays. The phone rang once at my mother's apartment in Schenectady, right in the middle of our Thanksgiving dinner, when the hourly workers had seen fit to destroy a time clock, and the supervisor had seen fit to make it *my* problem. Thank you, but no thank you.

Resist or no, though, and like all other salary personnel, I'm sent to corporate training in Milwaukee, the city made famous by another brewer. One week each of human relations/EEO training, problem solving training, and something called "Dimensional Management Training," or DMT.

The human relations/EEO seminar is pretty basic stuff—Maslow's hierarchy, satisfier/dissatisfier theories of wage, and so forth. A relaxing week, in all, and the company pays for it, to boot. And one of the general contractor bigwigs headquartered in this working-class town of corner taverns and brats takes me out to a Bucks game. A great time, especially for someone whose life orbit has been upstate New York.

The problem solving seminar—also known as the Kepner-Tregoe/Genco seminar—provides instruction in a set of highly organized methods and strategies designed to help employees to facilitate problems, potential problems, and decision-making—critical path planning, in essence.

Back at work after a week of problem solving training, I'm prepared to apply what I've learned. We've been alerted to a new corporate policy that affects the way we write up capital appropriation requests. So as I've been instructed at the seminar, I ask my boss to let me know what the overarching purpose is behind this policy.

Knowing the overarching purpose will presumably permit me to anticipate possible negative reaction, and to accommodate this in my writing—to bring my writing in line with audience expectations, hence to *control* reception—thereby increasing the probability that projects will proceed without delay.

After all, I'm not writing fiction or poetry, my aim is never to surprise or upset my audience. Hell no. Instead, I wish to secure the agreement of readers—readers with institutional power—to help them to understand that what I propose is worthwhile, and perhaps worthwhile to them, in some obvious way. So my primary aim as a writer is to reduce unwanted resistance, noise. I wish to enter fluently into a contract with my readers in which language will be rendered utterly transparent, in which words, repeatable 1-2-3 as numbers, will be made to speak directly to their customary corporate referents, and with not as much as a residual inkling that these selfsame words have a life of their own—

Joe, just *do* it this way, OK?

OK Tom.

DMT is the most challenging of the three training seminars. You find yourself thrown into a group of strangers, with a number of intensive role-playing tasks that require cooperation among members of the group. Tempers tend to flare as deadlines near, the fourth and final day of assigned tasks lasting well into the night. Any form of insurgency is all but eliminated by the sheer collective effort required to finish on time.

But the hidden agenda here has to do with industrial-organizational psychology, and the assigned tasks are in effect a ruse to draw out the "real you." On the final day, everyone is asked to evaluate, not the work accomplished, but one another. A four-quadrant grid is used to map personal qualities, warm/cold on the abscissa, dominant/submissive on the ordinate. Participants take turns shouting out adjectives that describe each monkey-in-the-middle, with a time clock to make it seem either a pressing exercise, or a game show. Each adjective is chalked into a given quadrant. I learn that I'm primarily a dominant/warm personality—excellent leadership material—so for me, it's a reassuring experience. Some participants leave that final session in tears.

When I return home, I wonder what, exactly, I'm to make of this newfound information, what sort of self-knowledge it represents. I can't see then that systems predicated on dominant competitors are designed to locate, reinforce, and reward . . . dominance. Just ask anyone in industry. Even in today's presumably more enlightened centers of manufacturing, it's far better to behave dominant/cold than submissive/warm. You believe otherwise, you're fucked.

It's a global strategy. They need young guys like me, to help corporate cooperation and teamwork "evolve"—competitively. Like the old-timers say in defense of U.S. military intervention in the Middle East: *we need that oil.* And for a young guy like me, toiling away in a technological domain that seems at once quotidian and gloriously reliable, it's tough to imagine an alternative to oil. At the same time, this sort of training *is* employee centered in one key sense: the company, however self-serving or misguided, cares enough to *pay* for our transformations.

Still, the real damage done by DMT is not fully evident until I run into Kevin.

Kevin is a real prick, through and through. So naturally, they send this prick Kevin to DMT, and post-DMT Kevin is an absolute wonder to behold. He's reformed, this bastard. He's a new man.

Kevin actually *requests* things now—he doesn't demand, he doesn't walk up and get in your face, he doesn't grab you by your lapels (if you have lapels) and threaten you with a beating. And he doesn't phone you to exchange mouthfuls of four-letter words.

No. Now Kevin the prick, armed now with the self-knowledge that people see right through him, has become poised Kevin, compatible Kevin, cunning Kevin. Now Kevin's less savory institutional survival instincts have been forced underground, only to resurface as deviously contrived plots, hatched to manipulate you into doing what he wants.

So, Joe, I'd like—so how are you?
OK, Kevin. What's up?
Well, I'd like to know if you'd be willing to help me out.
Help you out? OK—how?
Well. We have this problem over in the brewhouse—
Yeah?
—and—yeah—and since you're responsible for—
Me?
—yeah—I thought I'd ask you—

And if you should end up pointing out to Kevin that there is simply no way you can accommodate him because what he's asking for—as he knows only too goddamn well—is not within your operating budget, or his operating budget, or anyone's operating budget, Kevin will resort to intimidation of a different character.

If you're telling me that you can't do this—favor—

Yeah, that's what I'm telling you, Kevin. And you *know* I can't—we've been through this before. You need to budget for this sort of thing—we just don't have the funds for it.

—then I'll have to let your boss know how uncooperative you're being.

Have at, Kev. Right down the hall, first door on the left. But he won't say any different.

Me, I'd much rather deal with Kevin as I have in the past, after he walks up and plants his foot, hard, on the side of my desk.

OK pal—you and me, out in the fucking parking lot, NOW.

I grow a moustache, like Mike. My father teases me, calls me "whiskers." But facial hair is par for the course in upstate New York, especially during the winter months. When the shit is about to hit the fan, weather-wise, we men are expected to demonstrate the animal man cosmetically, and snow shovels in hand.

18 January 1980. The drive home from Fulton is just under twenty-three miles. On a good day, I can make it there in half an hour, most of it at 60 mph, thanks to Route 481. On a good day.

The town sits along the banks of the Oswego River, ten miles south-southeast of Oswego, which is right on Lake Ontario. Fulton is located at the edge of the snowbelt that bombards places slightly to the north and east, like Mexico, and the Tug Hill region even further northeast. An average winter in Fulton brings around 200 inches of snow.

Oswego County may be a poor county. Last time I checked, it had a relatively high rate of incest, for instance. Just drive around and you'll see how folks are struggling. But in Oswego County, they know how to plow snow. The trucks are out the moment it starts coming down, and most of the roads, country roads, experience light traffic, making it easier to plow.

Route 481 was built across a stretch of fairly flat, occasionally marshy land, with limited tree cover, numerous farms and a few gently sloping hills along the roadway. It runs north-northwest from Cicero, north of Syracuse, to Fulton, at its end rising over a grade and banking around to the right (northeast) and down, the brewery visible out the driver's side window.

When the wind blows and the snow falls, 481 is one of the worst roads in Central New York. The snow either blows directly into your windshield, or comes across the road so hard you can't see ten feet in front of you. Whiteout, as bad as anything you're likely to see in the continental U.S. outside of high-elevation mountain regions.

* * *

It's four o'clock on a Friday. The clerical staff workers, and anybody not directly related to production, have been told they can go home. Chris and I are wondering why they won't let *us* leave. Chris suspects it's a male thing—nearly all of the clerical staff are women. We're standing in the cafeteria, looking out the window at the guardhouse in the parking lot. It's a hundred yards away, and we can barely make it out.

Really coming down.

Yeah.

At five sharp, we put on our coats and step out the door of the main office building.

The wind and snow hit us with such ferocity that Chris's eyeglasses are half full of the white stuff in fifteen seconds. It's cold—not subzero cold, but as cold as it generally gets when this much shit falls from the sky. Chris takes off his specks, pushes the snow out with his pudgy fingers.

We're both smiling. Sort of.

See ya Monday. You be careful.

Yeah, you too.

The brewery plows are barely keeping up with the snowfall, and the parking lot lanes are two feet deep. I trudge up to my car. The snow is banking up over my headlights, blowing into and filling my front grill. I'm still driving the '71 Impala, piece of shit I picked up for three hundred bucks. The back quarter-panels are more Bondo now than metal. But it runs good—I've seen to that.

I open the door, a few inches of snow falling over my driver's seat.

Fuck.

But the crate starts up fine. I turn the defroster on high, grab my snow brush, and spend the next ten minutes using two hands to push and pull a foot and a half of snow off the top of my car. Then I scrape the thin layer of ice off the windows. Finally, I kick a pathway in front of the car, for a running start—which is really only a walking start. You need momentum to move in this stuff. It takes me twenty minutes in all, my fingers numb despite my gloves.

I get into the car, put it in drive, and start to ease out of my spot, spinning and sliding this way and that as I make it, slowly, out of the parking lot. I turn right, up Route 57 toward 481.

I take the right onto 481 and begin the slow climb up the hill.

Suddenly, before I'm even aware of it, I can't see a thing. Not a thing. I can't see my headlights glare. I can't see the middle of my hood.

I slow way, way down. I can't stop, because it's likely someone is behind me. My speedometer is bottoming out. I crawl forward, my foot as gentle on the accelerator as possible. The wind is buffeting the car to the left, out of the northwest, hard.

In another few moments, I've completely lost my bearings. Disoriented, I can't judge whether I'm headed for the guardrail that borders the right lane, or for the median on the left. I can tell I'm headed uphill only by the fact that my foot is instinctively bearing down a bit on the gas pedal to keep the car moving. I know that, a little further down the road, the median becomes a grassy area perhaps thirty feet wide, cupped in the middle for drainage. Cars have been known to skid off the road and into this median, rolling upside-down.

I open my driver's door, one hand on the wheel. I look down at the road to see if I can spot the lane markings. I spot a glimpse of yellow, turning the wheel to the right quickly, but not too quickly. I continue driving like this, looking down more than up, jerking the steering wheel a bit to the left, then a bit to the right, then a bit to the left, all the time modulating my gas pedal, ready to jerk my foot over to the brake pedal.

Minutes later—seven-eight minutes?—I begin to see the nose of my car, my headlights reflecting off the snow. Then the road itself begins to reappear. As I push through the end of the whiteout, the road ahead opens before me, white and calm. I step it up to 45 mph, and a half mile down the road, I look into my rearview to see the headlights of a semi gradually appearing out of the eerie white. More an off-white, more a grey.

It takes me an hour and a half to get home that night. On other snowy evenings, the plows will pile a huge mound of snow at the entrance to 481, preventing the more ambitious among us from testing our driving skills. I'm left wondering whether, when they built 481, the state contractors really gave any serious thought at all to the snowfall and wind directions in that region. I'm left wondering what they really knew of the region. I'm left wondering what they knew, and how they came to know it.

More engineering?

30 July 1981. Tom and Ted have stopped by my office after lunch. They begin to test their more evangelical impulses on me, explaining that rumors of a mysterious new epidemic are a clear sign that homosexuality is a sin, and that God (that's a capital G in their eyes) is therefore not to be doubted. Tom is especially clinical.

God says that if you stick your thing into somebody else's bum, you'll get what's coming to you.

Chris looks up from his desk, squinting. I can tell he's about to bust Tom's chops but good.

I wonder if Christ was a homosexual.

Tom and Ted look at him, more puzzled at Chris than ever. They get up and leave, ever so polite. Chris smiles, I'm laughing.

Sometimes I wish those two jerkoffs would take those silver fucking spoons out of their mouths and stick them up their born-again asses!

Chris laughs, hard, a rumbling laugh from deep in his gut.

They're OK, Joe.

Yeah, I know. But still.

Yknow—there's really only one feasible proof for the existence of god.

What's that?

That lakes and streams don't freeze solid during winter.

Huh?

Well, why don't they?

Because—because the layer of ice protects them from freezing?

C'mon Joe.

I struggle with Chris's question, but can't quite get at it.

OK, what's the reason then?

Maximum density of water occurs at—

At 39.2 degrees F. Yeah, OK, then—

Before this temperature, the colder water will continue to fall to lake bottom, and the entire lake will gradually cool—44 degrees, 43, and so on. But after 39 degrees or so, the colder water becomes *less* dense—and stays where it is, getting colder and colder until it freezes. Water's not your customary fluid that way—it crystallizes as it freezes.

So the proof of god is that ice is a crystalline structure?

Something like that.

Chris winks, I roll my eyes.

You going to the ISA meeting tomorrow night?

Oh I don't know, Chris.

I've been to a few such meetings, and I find them kinda, well, boring. But I don't want to hurt Chris's feelings. Most of the guys who attend the meetings are in their forties and fifties, and my interests don't seem to mesh well with theirs. Unlike Chris, they don't talk about anything but engineering. "Let's instrument that beer pitcher for low level warning." Or sports. "What about those Orangemen?" Or women. "Low target density in here." These guys can be amusing at times, but it's just not my idea of a fun evening.

Still, I decide to go.

I arrive at the Lakeshore Manor in Liverpool in my new used car, an orange '74 Nova, wearing a jacket, no tie. Walking into the small banquet room, I spot Chris, seated at a table with six other men—looks like the usual bunch. I take a seat next to Chris. The salesmen are already at it, schmoozing, drinks in hand. For them this is very much a business meeting, because they're socializing with potential customers. Something else I don't like, but it seems to be the way of the world—this world.

One guy at our table is especially talkative, and a bit loud. I haven't met this guy before. I lean over, mumble into Chris's left ear.

Chris, who's Mouthy?

Oh he's been around forever, sells control valves.

Oh.

The wait staff brings the dinner salads, in small faux wood bowls. I tune in to what Mouthy is saying. I don't think I like what I'm hearing.

Turns out Mouthy is telling dirty jokes. Now, there are dirty jokes and there are dirty jokes. At least, that's how I see it at the time. And most of Mouthy's are not the sorts of jokes you might get away with in mixed company.

Most of the table is laughing. Chris is laughing, too, but I can tell he's laughing as much *at* the guy as with him. Of course you have to know Chris to understand this. And Mouthy doesn't really know Chris, not like I know him.

He doesn't know me, either. But he goes on and on. As the wait staff, mostly women, bring the stuffed chicken breast, Mouthy is going off like a Gatling gun. He's shifted gears, fallen into a routine of question & punch line answer.

What's the difference between Italian women and Jewish women?

The question draws a blank from the table, Mouthy pausing just a moment for effect.

Italian women have *fake* jewels and *real* orgasms!

A round of laughter. *Stop right there*, I think to myself. I'm part Italian, which part is not upset in the least, but I see one of the waitresses wince. *Stop right there, don't let it go any further.*

I'm beginning to perspire. Mouthy continues.

And continues. And then he reaches that point that nobody should ever reach, that point that's reached every livelong day someplace on this despairing planet of ours.

* * *

You know the only thing worse than syphilis?
He pauses. The table is stumped.
Nigger syphilis!
He bursts into laughter, bellowing guffaws that can be heard two tables away. The wait staff pretend to ignore him, casting quick glances at one another. The other men laugh, some hard. Chris lets out a half laugh, squinting and tilting his head at Mouthy as if observing an odd specimen. I take a quick look around the restaurant, trying to be discreet—all whites of one shade or another.

But at this punch line, something happens to me. *In* me. I can feel my face flush, my gut tighten. I'm looking down, trembling a bit. I begin to raise my head, slowly, and as I do I begin to speak, calmly, a bit of a quaver in my voice. Trying to remain proportional, trying to maintain set point, trying not to be—not to be my father.

You know something—I don't think I like—what I'm hearing.
I'm looking right at Mouthy. The table goes silent.
What's that?
He's looking at me, jaw hanging, gaping.
I don't like what you're saying. It isn't right.
Would you marry one?
That's not the point, I—
Would you marry one?
I SAID that's not the point!
I pause a moment, trying to regain control.
I don't like your jokes—I don't think they're funny. I—don't think *you're* funny.
You like niggers huh—you like 'em?
Silence. I look at Chris, who looks down as he shakes his head in disgust. I too look down for a moment, searching myself for an answer. Then I look back up at Mouthy.
You know something?
What's that?
You know what?
What?
I think I know just enough—I think I've been around just long enough— to know that there's no point in my saying another word.

I turn my head away, Mouthy starts up again.
And I ignore him. For the rest of the evening I sit there, eat my dinner, chat with Chris. I just sit there, while Mouthy goes off.

I don't know, I can't know, I'll never know if I'm doing the right thing, the best thing.

But I do know that this is the last ISA meeting I'll attend. And without my telling him so, Chris knows this too. My friendship with Terrence will last until he too is born-again, five years later, but I'll never find the courage to tell him about that evening.

Mike quits his job at Xerox. He's got a new boss who wants to bring in his own men. And Mike's new boss has informed Mike that he might as well hit the road. You don't have to tell my brother twice.

So Mike leaves the Rochester suburbs, moves back to 501 Raphael Ave. for a short spell. Things are tight for a while, my paycheck carries the load. But within a few months, Mike lands a job with General Electric out on Farrell Road. My uncle Frank works there, is about to retire. A few months later, Mike will move out for the second time. The last time.

Eventually I'll land a senior engineering job in Syracuse, at a major pharmaceutical plant. There I'll meet and befriend a Jamaican-born engineer roughly my age, Terrence, whom I'll bring to an ISA meeting eventually.

Or at least, I think I do—I just can't recall, and I have no means at this point of verifying whether I did so. One *tries* to speak truth not simply to power, but to fact, to tell it like it was even while fighting shy of suggesting sheer verisimilitude. Events are telescoped, characters are characterized. After all, a book is not, can never be a life, so there is some inevitable faking entailed to render a readable narrative.

Still, one ought not to cover up all those nicks in the finish—better to let some persist as nicks of time, evidence of circumstance and urgency and intervention. A stressed finish? This is not my father's book.

So I sand with the grain, until it no longer figures, figuratively. Until I weary of my crafty self-control, and go back to the ordinary Joe I've always wanted to be. I can never be. Can never be, working with or against the grain. Taken with or without a grain.

Reflexive commentary, or reverse engineering: the end must be near.

These days I like to imagine Terrence attending all of the meetings I attend. I populate my meetings with his absence, with all of those absent. I look over my shoulder to ensure that everyone is present, every single motherfucker. I try to watch my mouth, take note of the damage it can do. Having given up on damage control, I figure the best one can hope for is after-the-fact repair. I salvage what I can of myself, of others.

People tell me these days—office people, as I work in offices, classrooms, and classrooms populated with future office workers—that I'm an outspoken son of a bitch, an angry malcontent. They tell me without telling me, and I hear them loud and clear. So I look over my shoulder, populating, try to watch my mouth, my presence and absence of mind, fret over blood relations, bloodlines, their relation to race relations. Does it matter, finally, how easily one bleeds, what's in one's blood?

They tell you it matters. They tell you without telling you.

O positive, and have been known, once in a great while, to wink. To ask for trouble.

One thing I *can* recall with certainty: introducing Terrence to Chris. The two men get along well that one evening, maybe a year after I leave the brewery.

But OK. My ethnicity is waning, so—

Before I leave: My brewery job has been my first job out of college, and I eventually realize that this means I'll never be treated as anything but a junior person. So if I want to accomplish what I'm expected to accomplish—to *advance* this somehow detachable thing called my *career*—I have to switch jobs. Even Tom concedes as much.

Don't know what to say, Joe. I hate to lose you.

So on a cool autumn Friday at five o'clock, Chris and I are left standing with each other in the brewery parking lot. I know I will see him again. And I know too that things will never be the same between us—we won't be working together, I won't be learning from him, day by day.

We shake hands. I can tell he wants to hug me, but something stops me from making that slight fall into his beefy forearms. The same something that prevents me from hugging my father, until it's too late. Too late to imagine any other end.

You're a good man, Joe. I'll miss you.

I'll miss you too, Chris.

Or something to that effect. It was all very real, as we were people, not personnel.

I take Route 57 home. Driving south along the river, the current is moving in the opposite direction. I wonder about endings and beginnings. I wonder why.

I don't know, can't know then that I'll find myself a couple of years later, at 4 pm, walking through the pharmaceutical plant, at one time the largest producer of penicillin in the U.S.—a major corporation that, like all major corporations, could do better by its employees, much better—distributing copies of my four-page-long letter of resignation to anybody and everybody who'll have one.

After, I'll arrive at Cindy's apartment on the north side, a bit shell-shocked.

Joe, what are you doing here so early?

But they'll never spend the money I say.
Just do the preliminary design and cost estimate he says.
But they'll never spend the money I say *I've been through this time and again.*
Just do it he says.

But I won't *just do it*. Not this time. No fucking way. I wasn't hired as a senior project engineer to do cost estimating, never to see a project through to completion. I can't be happy doing anyfuckingthing.

I can't be happy doing *this* thing. Can't—not any longer. No. Uh-uh. Not a chance. Just can't.

And I'll put the inappropriate question to the VP of Operations in a roomful of engineers—all men. I'll put that fucker on the spot, and he'll hedge.

So how much are you willing to spend to find out which department is using x amount of steam? I say.
Whatever it takes for 95% accuracy he replies.
$200,000? I say.
Whatever it takes he says.
$100,000? I say.
Whatever it takes he says.
Eyebrows are raised.

And not long after the eyebrows I'll receive a letter threatening me with disciplinary action, "up to and including termination"—for, as I see it, speaking out of turn.

So for 95% or better accuracy I'll design a vortex meter installation at a projected cost of $100,000 (±10%) to meter what amounts to 5% of the plant's steam flow, all because two operating areas refuse to consider a 50/50 split for accounting purposes. Metering steam flow might encourage a given operating area to conserve, true, but the steam to be metered would nonetheless remain a small percentage of the plant total.

So steam metering, by virtue of my instrument and control expertise, will first and foremost provide information as to *who* is using *what*—metering *qua* monitoring. It's all internal accounting, approximately the same amount of coal or fuel oil or natural gas entering the boiler house during a given fiscal period.

And I'll present my initial design and cost estimate at yet another staff meeting—all men—two months later. And the VP of Operations will hedge—again.

$100,000? he replies.
Yeah, $100,000—to monitor no more than 5% of plant steam consumption I say.

Well then why can't we get the two operating areas to agree to, say, a 60/40 split? he says.

And my boss will immediately interject, arguing that we need to meter. And the VP will concede.

And eight months later they'll decide, finally, *not* to pursue the project. As I knew they'd ultimately decide. Me, senior cost estimator. Me, senior straw boss of the fucking whiz-bang calculator. Me, wet middle finger stuck in the institutional wind.

But that first piece of paperwork, that disciplinary letter, will long since have been securely tucked away in my personnel (not person) file. Step 1 in eliminating an employee: you create a paper trail. And it always helps if said employee doesn't get along too well with his boss, with his boss's boss. Is an outspoken son of a bitch.

I can take a hint, like my brother, but the company's timing will be just a bit ahead of mine—three months, to be exact, which I'll ponder as they escort me out the door, my chest still congested from yet another bout of bronchitis.

They fired me I tell Cindy.

15 March 1980. Mike and I are leaning our elbows on the rail that girds the third-floor landing of our hotel in San Diego. We're looking out over the city. Stan and Greg are taking showers, Rick's next. It's seven in the morning, the sun just above the hills to the east, a glint of morning dew in the air. Traffic is just picking up. The palm trees look like still lifes. A couple of sailors stroll by below us wearing monkey suits. We've just phoned back home, and my father has assured us that we'll be returning to snow. I shake my head, turning to Mike, wondering what it means to be a rock, and not to roll.

Mike, what the fuck are we doing in Syracuse?

I don't know.

But we both know then what we're both thinking—that town don't look good, in snow.

8 December 1980. Ima—

4 July 1982. And what about Nora?—that wickedass accident she had on Thompson Road with that jerk who rolled the jeep, the fireworks we enjoyed together at Cazenovia Lake, her son on my shoulder, and—but this isn't the time to go there, endings mounting. Besides, I can feel my ethnicity about to expire.

20 June 1977. When I drive up to Fulton to interview at the brewery, I ask the standard question all new graduates are counseled to ask about upward mobility. The personnel rep pauses, evidently weighing my question carefully, measuring his response. Then he motions toward the ceiling.

The sky's the limit.

The sky's the limit.

Five years later I'll receive my professional engineer's license. Once an engineer, always an engineer.

Say it ain't so.

Eight years later, I'll be sitting in my mother's apartment in Schenectady, alone, when I notice a Corvette convertible pull up. A striking, six-foot-tall blonde will get out and enter our building. Next thing I know, I'll hear a knock at the door.

Hello. Joe Amato?

She'll be posing, poised, her left hand behind her back.

Uh, yeah.

This is for you.

She'll bring her hand around, and hand me a folded document. A subpoena.

One of the laborers who worked at the brewery has sued the brewing company for a million bucks, claiming that he'd slipped and fallen on the construction site due to inept snow removal. And the brewing company has sued Bruce. So Bruce's attorney has subpoenaed me to testify that snow removal was indeed handled properly.

Of course it was. Pushing a wheelbarrow over a construction site midwinter is bound to entail certain occupational hazards, like any job.

I'll feel sorry for the guy who fucked up his back.

I'll chat for a while with the woman who's served me, a devout Catholic who doesn't believe in birth control. I'll try unsuccessfully to argue her into a more sensible position. No luck.

A few weeks later I'll receive a phone call from Bruce. He tells me that the brewing company has dropped the suit against him.

Wait—she was a brunette, not a blonde.

Eleven years later, after a record 741 weeks straight, Pink Floyd's *The Dark Side of the Moon* will finally fall off of *Billboard*'s "Top 200 Albums" chart.

Twenty years later,

Twenty years later,

                                                                                            & skyward rise

            as the markets begin their skyward   & fall

the brewery will shut down.

    And a few years into the next century, the oldest chocolate factory in the U.S., a mile or so downriver from the brewery, will also close its doors, even as GM execs set about planning for the last Olds to roll off the assembly line.

    And in another few years, the brewery will be retrofitted to become the Northeast's first ethanol plant. Ethanol from corn initially, then from the by-product of paper production at Ticonderoga's International Paper plant. Paper from willow tree chips, mostly. That's the plan, at least. Wood to fuel.

    One hundred plant personnel, or about one-twelfth of the former brewery workforce.

More engineering.

NB   It's all earth science, and earth science is at its core a social art—the collective primate's best guess on a given day. Technology as habitation, yes, but habitation as the species infrastructure, inside out. Turning outside in, serving and shaping and recycling the us we are ever in the process of becoming. Ritual to logic to habit to plot to punch list, its more elemental effects found in the flotsam of practice, the workable rot and contrived junk of thought discharged. Outdoors, under the stars, become fertilizer, shit, worm feed, or compound structure. More rot, more junk, but no less profound than opposable thumbs. Like gangrenous tissue, or dead atoms, like alloy, or DNA, like keyboard, or blueprint, like letter, or wave, or oxide—of and out of organic and inorganic our reworkings of body and mind and hopeful calibration of spirit: it's all a contribution to the meandering real. Composing and decomposing, out of this compost life again arising—

There is no plan, no plan to be unearthed. Retrospect reveals only rhythms of flesh and blood, grains of truth, rhymes of passion.

My primary objective in pursuing the doctorate is to fully develop my potential as a writer. I believe that a thorough understanding of literature is essential to one's creative insight. I am most interested in enhancing my abilities as a poet, but hope to engage in all facets of writing, including fiction and journalism. Ultimately, I hope to share what insight I have acquired by helping others develop their writing abilities and by exploring new ways of interpreting literature. Teaching and writing, the way I see it, will become the foremost concerns of my professional life.

My past work experiences in industry have been responsible for convincing me of the need to pursue a more creative and scholarly career. Inasmuch as I feel secure when dealing with technical concepts, working as an engineer for the past seven years has been an invaluable aid in coping with an increasingly technical world. However, I have become aware of the fact that I differ from my colleagues in terms of my interests and hobbies, and thus my perspective on living is at times at odds with theirs.

In order for me to function in an engineering environment, then, I have found it necessary to subdue my creative interests. I have done so for nearly seven years now, largely for financial reasons. In particular, I have supported my father for the past four years. He has now reached retirement age, and I therefore feel free to finally direct my efforts toward what I believe to be my true vocation.

I hope someday to be able to elucidate my experiences in industry by making them the subject of my writing.

9 April 1984

foraging for redolent redaction

he seeks the archived { I }   a gangling text   of research into

                                                               some self's infractions

       to yield a cheese redundant as the next

cheddar: mushrooming consumption exhumes

all retractions, correspondence works through

correspondence to winnow, then subsume

all trace of golden silence;   words in spew

the task begins to take its final

                                        shape

masquerading as the facts of sheer truth the truths of mere fact manufacture
    faith

              to read the sum as   a someone   in lieu . . .

the eye that takes the measure of such stock

                 would do well to consider what is   knot

Epilogue
Variable Cloudiness, Chance of Precipitation 50%

>The engineer simply cannot deny that he did it. If his works do not work, he is damned.
>
>—Herbert Hoover

—across the distances

What keeps us going

You know how it goes—beginnings, middles, endings, ends.

Sunlit desolation, grey-white cloudbanks, empty with beauty. Far as the eye can see.

Always just before, or just ahead. Always hard on our heels. But inside or outside, not, never can be *here, now*, precisely.

Awareness. An act of faith? I'll never be sure. I mean, what do I know.

A moment later, gone.

Now, just me and my old man.

Once in a while, instead of Abbott & Costello, we watch the Sunday morning news shows together. Right around the time *supply-side* falls from the lips of those who presume to face the nation, we start having water problems. Again.

This time it's low water pressure. The landlord won't budge, and Freddie calls the County Board of Health for the second time in nearly twelve years. They come to inspect, and condemn the place. Tell the landlord either to repair the water supply piping, or stop collecting rent.

We get the eviction notice, signed by Harris. We've got one month.

Have I said this already?—there are greys and there are grays. In early fall, splotches of yellow, orange, red are fired by the bleak autumn sky, no small consolation for the earthbound onslaught in the offing. But as spring approaches and the snows taper off, the off-whites and browns would fuse with the clouds, become a mishmash of uncertain hue. Summer might be a triumph, might not—depends on the summer. Or on what you make of a partial reprieve. On your way with words.

That's right, the easy way. Get the fuck out. Move. Put it behind you. Pull yourselves together, and up. Both of you.
Both of you.

Freddie and his family are the first to move.
I run into Freddie a final time, a decade later, at the bank. He's gone grey fast, I've got a grey hair or two myself. Somehow I can tell he's genuinely happy to see me, and somehow I'm genuinely happy to see him. He's finally getting a loan to buy a house.
That's great Freddie. You still driving for the schools?
Yeah. It's going OK. How's your dad anyway?
He's doing pretty good.
That's good. Cripes, tell him I said hello.

A few months after he receives Joe Amato's Christmas card—the last he'll hear from the Joe Amato he met during the war—my father will find himself in a two-bedroom apartment in the village, across from the Liverpool drive-in, living with his oldest son. Newish complex. Washers and dryers downstairs, a tiny storage area in the basement. Air-conditioned, wall-to-wall carpeting, dishwasher, electric range. Natural gas fireplace built into the paneled walls. We even order HBO.
Pretty snazzy place, Joe.

My boss Tom is worried. He sees me as a guy with no motorcycle, no car payments, no kids—no commitments. He knows I live with my father. But that's all he knows, because that's all he needs to know, and that's all I tell him. Walking from the parking lot into work one rainy morning, he expresses his concern.
We gotta find a way to keep you wanting that paycheck.
Uh-huh.
I end up buying my first new car. My first four-cylinder. My first car with front-wheel drive. My first foreign car—a small Datsun 310GX.
Well, not new exactly—it's a demo. But it's in perfect shape.
My father is impressed with it. And this way I'll give him the Nova, we'll each have a car.

Epilogue

* * *

I get my father to quit smoking and drinking, start exercising. After three months, he looks great. But with no full-time job in the works after four months of looking, he begins to falter. Again.

I talk to my mother. I'm depressed, don't know what to do.

Mom—I need to tell him I need my own place.

Joey, you just have to say it.

I know.

Living with your father at 26—can you imagine? You have to live your *own* life, Joey.

It's another six months before I muster the resolve to pop the question, after supper one night.

Dad—

Yeah Joe.

I think it's time . . . you found your own place.

He hesitates for just a moment, a lifetime.

OK.

Don't worry—I'll pay for it. It's just that I think it's time I got my own place too, yknow? And besides, maybe I'll get that job in Saudi Arabia, make some *real* money.

Wouldn't that be something—you deserve it.

He's not looking at me, I'm not looking at him. By now I know the job working for ARAMCO in Saudi Arabia won't happen. He knows it too. Having grown more aware of my cultural distance from the Middle East, I'm not so sure this is a bad thing. But that's not the point.

And listen—I want you to have the TV.

No—you should keep it, Joe. It's a beautiful set.

No, Dad—I have the stereo.

Helluva stereo you bought.

Yeah. So you take the TV set, OK?

OK Joe.

Craft: originally, etymologically, suggesting a notion akin to "strength," a word underwriting skill, trade, profession. Collecting forces, collectives of quantum energies, crafting hands and crafty heads. You can witness it all in the classifieds.

It's like this: workforce = work + force. Work = force through a distance. Forced through a distance, distance forced?

I can say it now: formulae for those who haven't learned to read the writing.

Still, we're all a little slow on the uptake at times. Laborers to a one, we've earned it, right?

: He's looking at eighty bucks a month pension from General Electric, for nearly twenty years of work, because he took his severance pay. Maybe another five hundred a month from Social Security.

Can he see it coming, this penultimate split?

Don't shit yourself.

Driving down Buckley Road toward Bailey, I pull off and park on the shoulder sometimes, walk a dozen steps into Dixon's large, tree-studded lot to have a long look at 112 South Dolores Terrace—from the back. Like most ranch houses, the front of the house has a picture window, and I'm trying to be discreet. But truth is, I feel entitled, front *or* back.

The house is still white—white aluminum siding now. The locust tree that served as first base, the maple tree at the base of which Cognac is buried, all the trees my grampa gave us are much larger. Like the evergreens out front, larger than my folks ever imagined they would be. I'm larger too, so the backyard my father would always brag about as being *deep* seems just a bit small.

The people who live there now still use a clothesline, like we used to. My mother would run our clothes through the wringer-washer, and then hang them on the line, weather permitting. I'd help her take them down.

Over time my grampa's trees will grow too big, and the new owners will have all but one removed.

Every now and then Mike and I drive over to Raphael Ave., just to check out the place.

One day, years later, it's gone.

According to the old lady who, every morning and every evening, can be seen walking her little dog up and down the street, they hooked up some chains to a dozer, and it just came falling down. The mailbox post still serves as a marker.

Another year or so, and not a trace remains. No ruins to piss on, or to build a temple over.

I am a bit of a numerologist, have grown fond of that passing, pulsating harmony, 18436572. I stand in wonder of primes especially, especially that lucky number 7 (haven't you noticed?). GL4-2813, 457-6737: burned in so, these seem prime, letter to number to tone. But numerically they are not.

Epilogue

* * *

Past or in our primes, can we ever know? Even numbers are more symmetrical, but some say there is luck, even divinity in odd numbers. 1.1.3111 is a given, but life along the Erie Canal remains a struggle, awaiting further demonstration, documentation, even the sum of primes. $144 = 13 + 131$, and there are ten other such tallies that yield a dozen dozens. Those Prussians. And those indigenous peoples. Five Nations, one nation, ratio, ration, rationale. Etc. A thicket of intention in a haze of overlapping histories.

And so, as such stories go, I eventually make it back. Back to a place I have never really known, from which I could never have begun. But back nonetheless, to middle-class registers of conflict and need. It will be years before I learn, seasoned, that it is possible to discover winter, crystalline and exact, and know it well. Know it as anything but a design.

It's clearing now, just for a moment: there are no reasons, or there is a reason for everything, for nothing is meant not to be. Times change, from joy to despair, and back. If you're lucky. If you're lucky, you struggle on, on through the solitary luxury of failure.

The wind changes. A light grey, so light a grey approaching. Front moving in, difficult to see through. Find a way through.

I stumble upon myself in the cold.

A corroded line cuts off the flow of dreams, dribbles

away

Acknowledgments

THOSE WHO CREATE owe a debt to others who create. So thanks, first, to those others who create.

Anyone who's had the opportunity to swap ideas with Don Byrd is aware of his remarkably fecund intellect. I am absurdly fortunate to have encountered Don's instructional guidance upon embarking on graduate studies, and I learned much, too, from Don's colleagues in the Department of English at the University at Albany—SUNY. Thanks to Gail E. Hawisher, who solicited a piece from me some years ago on literacy and technology that led ultimately to my first attempts at memoir. Next in line is (the poet) Andrew Levy, who counseled me, wisely, not to worry too much about artifice initially and to concentrate on *telling the stories*. David Porush's sage advice—"No matter what, keep writing"—and Michael Joyce's unwavering support and creative precedent kept me going when things got tough. Without Greg Hewett's careful reading of the work in manuscript and judicious recommendations, what you hold in your hands would be considerably less than what it is. I'm deeply indebted to Jennifer Dorn, (series editor) Fred Gardaphe, and David Hamilton, each of whom published a chapter of this work in their respective journals. My gratitude to James Peltz for pursuing this project even when my own belief in it began to falter, and thanks to Gary H. Dunham and his staff at SUNY Press for bringing it to fruition. And I owe a special thanks to Zachary Karmen, with the Legal Division of Onondaga County Department of Social Services, for his permission to access my family's public assistance and Medicaid records.

Without the many relatives, friends, and acquaintances that appear throughout these pages, there would be no book. Thanks to my parents' parents, all of whom are gone now, in whose eyes my young eyes saw the pride of succession: my oma Johanna (née Bentz) and Henri Bourgoin (who died before I was born, a man of whom my father always spoke so highly); and Antoinette (née DePasquale) and Rosario Amato. Thanks to my aunt and uncle on my mother's side, Ilse and (the late) Eric Mariage, and to their family, who together imparted a sense of a world more expansive than the world I would otherwise have known. And thanks to my aunts and uncles on my father's side—Maryetta Tridente and (the late) Sam Amato; Mary and (the late) Frank Amato; and Dominick and (the late) Dorothy Amato, and to their family—whose examples have amounted, all told, to a lesson in family perseverance.

Thanks to my old friends for being old friends.

John M. Mishko, another great teacher, showed me how to be an engineer in the trenches—and so much more, to which his thinly veiled presence in this work will attest—and I learned from his wife, (the late) Bernice W. Mishko, that even engineers need an emotional life. I am ever grateful to these two kindred spirits, and to other kind souls too numerous to mention—coworkers, colleagues, correspondents, and students, past and present—for their generosity, wisdom, and love.

My mother and father remain, in death as in life, a constant source of inspiration. Enduring thanks, finally, to my brother Mike, who was there, and to my wife, Kass Fleisher, who taught me how to write a sentence.

Grateful acknowledgment is made to the following journals, in which earlier versions of several chapters originally appeared: chapter 7, *VIA—Voices in Italian Americana* 10.1 (Spring 1999); chapter 8, *The Iowa Review* 34.2 (Fall 2004); chapter 9, *Square One* 1 (Spring 2003).